WAITING FOR CONTACT

UNIVERSITY PRESS OF FLORIDA

Florida A&M University, Tallahassee
Florida Atlantic University, Boca Raton
Florida Gulf Coast University, Ft. Myers
Florida International University, Miami
Florida State University, Tallahassee
New College of Florida, Sarasota
University of Central Florida, Orlando
University of Florida, Gainesville
University of North Florida, Jacksonville
University of South Florida, Tampa
University of West Florida, Pensacola

WAITING

FOR

CONTACT

The Search for
Extraterrestrial Intelligence

LAWRENCE SQUERI

UNIVERSITY PRESS OF FLORIDA

Gainesville Tallahassee Tampa

Boca Raton Pensacola Orlando

Miami Jacksonville Ft. Myers Sarasota

This book may be available in an electronic edition.

21 20 19 18 17 16   6 5 4 3 2 1

A record of cataloging-in-publication data is available from the Library of Congress.
ISBN 978-0-8130-6214-3

The University Press of Florida is the scholarly publishing agency for the State
University System of Florida, comprising Florida A&M University, Florida Atlantic
University, Florida Gulf Coast University, Florida International University, Florida
State University, New College of Florida, University of Central Florida, University of
Florida, University of North Florida, University of South Florida, and University of
West Florida.

University Press of Florida
15 Northwest 15th Street
Gainesville, FL 32611-2079
http://www.upf.com

book subvention provided by
Figure Foundation
decoding the inward distance

CONTENTS

PREFACE

Going to the movies when I was growing up in Astoria, New York, was always a special treat, especially if a good monster flick was the main feature. Even better than giant ants running amok were aliens from some unknown planet in distant space. Somehow they managed to land on Earth and preach to us benighted Earthlings, or in a really good movie they wreaked havoc on us backward, defenseless humans. It was so much fun, especially because it was all make-believe—or was it?

The media reported sightings of flying saucers. Ordinary citizens came forward and told of meeting beings from distant worlds. What is one to believe? I could have devoted my life to examining these tales. I also realized that the quest would not pay the bills. I was drawn to something conventional: university teaching and the study of history, the focus of my doctorate from the University of Pennsylvania.

At some point in my life—I do not remember when—I became aware that a number of astronomers were searching for extraterrestrial intelligence. Although scattered around the world, these astronomers were mostly American and their common effort is known as SETI, the Search for Extraterrestrial Intelligence. Curious about extraterrestrials, the SETI astronomers were hoping that extraterrestrials were also curious. They aimed radio telescopes to the heavens, hoping that extraterrestrials were beaming radio messages to other worlds, and perhaps to ours. Although wrapped in modern technology, their curiosity is as old as human history. I soon realized, though, that the SETI quest is only partially the satisfaction of curiosity. SETI is multidimensional, a search for answers to questions humans have asked over the ages.

SETI excited me, and this work is the result.

A few acknowledgments are in order. My loving wife, Linda, has shown much patience during my research and writing. East Stroudsburg University awarded me two sabbaticals that gave me much leisure time for reading and research. The librarians at the university's Kemp Library did their usual professional job in assisting me. My research assistant, Monika Ackerman, was very efficient and enthusiastic. Jason Steinbrecher also helped me in my research.

I would also like to acknowledge the late John Billingham, who graciously shared his SETI memories. Special thanks also go to Gerrit Verschuur, who answered queries and sent me chapters from his SETI manuscript, which he published in 2015. I also thank Seth Shostak and Douglas Vakoch of the SETI Institute for answering my questions.

An earlier version of some of the material in this book was originally published in the *Journal of Popular Culture* (2004).

WAITING FOR CONTACT

Introduction

Imagine that ET has called. He or she (assuming ET has a sex) has sent a radio message to Earth.

The astronomers and astrophysicists who have searched for extraterrestrials are elated. They have aimed radio telescopes to the stars for over fifty years, hoping to catch an alien signal. They called their quest the Search for Extraterrestrial Intelligence, or SETI. Finally, the search has ended.

Basking in their new fame, SETI scientists assure the world that the greatest event in human history has taken place. Humanity will gain knowledge and wisdom from advanced aliens, make enormous leaps in self-understanding, and perhaps save itself from its suicidal tendencies.

Why do SETI people insist that humanity gains from contacting extraterrestrials? Why believe that extraterrestrials are benevolent in the first place? Science fiction has presented numerous scenarios of evil, genocidal extraterrestrials. Is not one of these scenarios just as likely as that of the good extraterrestrial?

SETI people are optimists. They know the universe was eight billion years old when our solar system formed. They posit older worlds in our galaxy, other worlds where, as on Earth, life emerged, grew intelligent, and survived. If older extraterrestrial races had not learned to cope with lethal technologies and live in harmony, they would be extinct. ET is a survivor, living in worlds that are older, advanced, and peaceful, or so goes the SETI narrative. Carl Sagan (1934–96) went a step further.

Noting that Earth shares the same elements as the entire universe, he reasoned that the progression from inanimate matter to life and intelligence was common and that our galaxy alone contains a million advanced races. Sagan was the supreme optimist.[1]

But ET has not called, and the SETI community still waits.

The SETI movement was born in 1959 when Giuseppe Cocconi (1914–2008) and Philip Morrison (1915–2005) of Cornell University published their seminal article in *Nature,* "Searching for Interstellar Communications." There they noted that the recently invented radio telescope could detect extraterrestrial radio signals. Whereas previous generations could only speculate on the existence of extraterrestrials, mid-twentieth-century humans could be proactive.[2] The following year, Frank Drake (b. 1930) used the radio telescope of the Green Bank Observatory in West Virginia to search two nearby stars. This, the first SETI search, Drake called Project Ozma.[3]

In the following decades, SETI searches were haphazard, for full-time SETI workers have been few. After all, if ET is not heard from, SETI scientists have nothing to show for their labors. As a result, SETI scientists have spent most of their professional careers in non-SETI pursuits. Drake is noted for his measurements of Jupiter's upper atmosphere. He was also director of the National Astronomy and Ionosphere Center, the formal name for Puerto Rico's Arecibo Observatory, the world's largest radio telescope. Morrison was a pioneer in the study of gamma rays and cosmic rays.

Access to radio telescopes also bedevils SETI researchers. Science wars can be vicious. SETI astronomers must compete with traditional astronomers for telescope time. Only since 2007 have SETI enthusiasts had their own dedicated radio telescope, the Allen Telescope Array, in California. Yet SETI's problems have continued, as that telescope has suffered from funding shortages.

If nothing else, SETI apologists have excuses for not contacting extraterrestrials. They invariably note the small number of SETI searches. They add that instead of microwave, extraterrestrials may be using other parts of the electromagnetic spectrum, or perhaps are communicating in ways we cannot imagine. After all, if Earth had received extraterrestrial radio signals before the twentieth century, the smartest

Earthlings would have been clueless. Radio was beyond the genius of Isaac Newton.

Astronomers from around the world have done SETI research. During the Cold War, the Soviet government sponsored a SETI program. Important work has been done at Harvard University and at the University of California, Berkeley (UC Berkeley). For a while, the locus of American SETI was in NASA, but Congress killed that project in 1993. Since then, a leader in SETI research has been the private SETI Institute of Mountain View, California, which owns and runs the Allen Telescope Array. In July 2015, Russian billionaire Yuri Milner (b. 1961) announced a $100 million donation to advance SETI research. Milner has spread his largesse, but the center of this new search will be UC Berkeley.

Over the years, the SETI community has debated search strategies, asking which radio frequencies ET is using and where in the sky to aim their radio telescopes. Along the way, SETI astronomers have vastly improved their equipment. Drake used a single-channel receiver in his first search; today's SETI uses multi-channel receivers whose sophistication rivals the jump in technology from Columbus's ships to the space shuttle. The fact that such technology is available, however, is not reason enough for a society to initiate an extraterrestrial search. Benefits must be seen before resources are expended. Convincing both government and private donors of a big payoff has been a perennial challenge to the SETI community.

Simply defined, SETI has been the use of radio to locate intelligent extraterrestrials and have a conversation. On a higher level, SETI has been variously defined as a pseudoscience, a biophysical cosmology, a modern myth. Each of these definitions has some validity.[4] SETI enthusiasts, though, regard themselves as explorers. They identify with Columbus, who discovered worlds that his contemporaries had never imagined.[5] That NASA chose Columbus Day (1992) to launch its first SETI search was no coincidence. Unlike Columbus, SETI researchers will not personally visit distant places. Explorers of a modern kind, they will use radio to learn of new worlds and their inhabitants.

SETI can also be seen as an offbeat political movement that avoids the traditional activism of organizing and proselytizing. According to

SETI promoters, contact with extraterrestrials will enhance humanity. In all, the SETI quest satisfies curiosity concerning extraterrestrials and as a bonus it might save the planet. SETI enthusiasts believe that theirs will be the most important enterprise in human history, once contact is made.[6]

In ways that the SETI community might not imagine or care to admit, Columbus is a fitting precursor. Like the SETI people, he had a political agenda. Columbus spent years in the courts of Europe, hawking his project for a westward voyage. He made extravagant claims for what he found. SETI activists have also lobbied, having approached the academic community, NASA, and philanthropists. Along the way, they have extolled the benefits of contact, ranging from cancer cures to advice for world peace. Politics obsessed several of the SETI grandees. Sagan and Morrison were lifelong progressives. Sagan's high hopes for contact's benefits were almost a default alternative in case political activism should fail.

SETI people are self-defined seekers; they have placed themselves in the continuum of the explorers of old, although in spirit they share little. Columbus, Cortés, the whole tribe of Spanish conquistadores, as well as the French and English explorers, were not looking for wisdom and guidance from other civilizations. They wanted to exploit. The conquistadores slew native warriors, raped their women, and stole their gold, all the while certain that Jesus was smiling. If European explorers had an agenda for the indigenous peoples after crushing them, it was to obliterate their culture and impose European ways. In contrast, the SETI people approach extraterrestrials as suppliants, hoping for wisdom and the deeper meaning of existence.

Columbus never realized the consequences of his discovery. He initiated the so-called Columbian Exchange, in which native crops such as corn, potatoes, peanuts, and many others, a veritable cornucopia of foods, forever enhanced the world diet; in return, Old World crops and farm animals were introduced to the Americas and greatly increased the world food supply. Gold and silver from the Americas in the short run enriched Spain and in the long run achieved a revenge of sorts by causing inflation in Spain, and then worldwide, rocking the economies of Asia and Europe. As for the indigenous peoples of the Americas, millions died from European diseases, and the survivors were forced to adopt new lifestyles and religions. If history shows anything, it shows that the future unfolds in ways that contemporaries never imagine. The SETI community is as much in the dark as Columbus was, in spite of its hopes that contact will better humanity. The SETI people may be a bright lot, yet not all observers share their optimism.

In spite of SETI's high hopes, extraterrestrials have thus far evaded detection. Life of any kind, even primitive bacteria, has yet to be found beyond Earth. Nonetheless, this absence has not prevented speculation about extraterrestrials, often with not-so-hidden motives. Whether in philosophy, literature, or science, the extraterrestrial has been seen as superior, almost utopian, to remind humans of their selfish, backward ways.

If SETI scientists hear from extraterrestrials, SETI's leading figures will be famous. If SETI fails to find intelligent life and humanity abandons the quest, SETI will be recalled as a historical oddity. Regardless of the outcome, SETI was a cultural phenomenon of the second half of the twentieth century and continues in the new millennium. It allowed leading scientists to express political anxieties, and it has always sought new answers to old questions. SETI has put old wine in new bottles.

1

Before SETI

The Age of Pluralism

Lost in time, beyond knowledge, is that moment when a human creature first looked at the heavens, saw the lights of the sky, and asked
if other beings lived in worlds above. We will never know when and
where a human first asked that question. We will never know if a *Homo
erectus* first asked that question or if a Neanderthal or Cro-Magnon of
later eras did the asking.

We know that people of early historic times believed in gods, mighty
beings who played a role in human affairs. The ancient Mesopotamians
(in today's Iraq) identified their gods with the planets to which they
ascribed great power over humans. Other ancient peoples believed
their gods lived in the heavens. Ra of the ancient Egyptians, Zeus of
the Greeks, and Odin of the Germanic tribes all lived in the sky, ignoring humans but when bored showing a keen interest in humanity.
Although extraterrestrials of sorts, the gods of antiquity were too much
part of this world. They are not the extraterrestrials of today, distant
and probably ignorant of our existence.

Greek Pluralism

The idea that other worlds exist beyond Earth and may contain intelligent creatures is known as pluralism. The classical Greeks, who had
an opinion on nearly everything, probably invented it. Although their

speculations on extraterrestrial life could be whimsical, some of the Greeks realized the greater use of pluralism. Extraterrestrials could serve as an allegory to comment on humanity. Although no one cared to flatter the human race by imagining inferior, villainous extraterrestrials, the possibility of using superior extraterrestrials to deflate swollen human egos was irresistible. The Pythagoreans of the fifth century B.C., best known for their belief in the centrality of numbers, claimed the moon supported creatures of great size and beauty, a life so advanced that it did not produce excrement.

Also agenda-driven were the atomists, the best known of the Greek pluralists, who flourished from roughly 480 to 280 B.C. The atomists theorized that all matter reduced to indestructible atoms, perpetually and randomly in motion. When atoms combine, they make up the stuff of Earth. When they combine beyond Earth, they create other inhabited worlds. For the atomists, extraterrestrials were part of a vision of life that encompassed the entire universe. Yet pluralism was secondary to a social agenda. The atomists preached hedonism, holding that the random motion of atoms destroyed purpose in life and as a result people should live only for the moment. Besides, death and the possible punishments of the afterlife were not to be feared, because there was no immortality. The random motion of atoms saw to that.[1]

The atomic theory had a modern ring. The atomists made a case for extraterrestrials similar to the SETI community's in that they argued that Earth's physical laws were universal and led to the same results. If life appeared on Earth, it should appear elsewhere. In addition, like many moderns who reject traditional faiths, the atomists could not accept an empty universe. Although they posited a random system of matter, they gave a purpose to the rest of the universe by endowing it with life.

The atomists did not represent the classical mainstream. That honor belonged to the polymath Aristotle (384–322 B.C.), whose common sense often gave him great insight into a host of subjects. In one instance, though, Aristotle's common sense failed him. He believed the sun moved over a stationary Earth. After all, unless one is in love, one does not feel Earth move. The ever-practical Aristotle went a step further and assumed that Earth is the center of all motion. A cosmos with only one center implies the nonexistence of other worlds. If they did exist, they would be sucked into the center, which is Earth.

Medieval and Early Modern Pluralism

Aristotle was the premier savant of the ancient world, and his prestige persisted during the Middle Ages (roughly A.D. 500 to 1400). The Catholic Church liked his rejection of pluralism, since pluralism implied that humanity was not the center of creation. Besides, extraterrestrials were irrelevant for salvation. There was a flaw, though, to Aristotle's geocentrism; it implied limits to God's power. Hence, Thomas Aquinas (1225–74) and William of Ockham (1288–1348), both leading philosophers, declared other worlds possible, although probably nonexistent.[2]

The possible became certain when the Polish astronomer Nicolaus Copernicus (1473–1543) forever changed our view of the heavens. By holding that the sun was at the center of the cosmos and that Earth moved instead of the sun, Copernicus created a new astronomical theory with astounding implications: he unwittingly questioned humanity's special place in the universe. The belief in human uniqueness gradually lost ground.[3]

Although Copernicus did not mention pluralism, he implied it. The seventeenth and eighteenth centuries would be its golden age. Scholars throughout Europe debated whether other worlds were inhabited and, if so, whether extraterrestrials professed Christianity, another religion, or none. Another pressing matter of discussion was the intelligence and ethics of extraterrestrials. To prick human egos, pluralists often posited superior extraterrestrials.[4]

During pluralism's golden age, astronomical information depended on primitive telescopes. By today's standards, much was unknown, giving much leeway to metaphysical speculation, often involving analogical and teleological reasoning. According to the argument from analogy, if two objects share many properties, that is, if they share properties A, B, C, and D, then if the first object contains E, chances are very good that the second object also contains E. The astronomer Johannes Kepler (1571–1630) used this reasoning to populate the moon. Since both Earth and the moon contained craters and mountains, Kepler deduced that the moon also contained living creatures. According to the principle of teleology, matter and life has a purpose, making a lifeless universe meaningless. Pluralists concluded that other worlds contained life.[5]

Besides sowing doubts about Christianity, pluralism also served another agenda: comparing humans to their extraterrestrial counterparts. These comparisons call to mind utopian writers such as Thomas More (1478–1535), who compared ordinary humans to idealized humans,[6] or cynics such as Jonathan Swift (1667–1745), whose *Gulliver's Travels* unfavorably compared humans to horses. Also coming to mind is the practice of pointing to the alleged failings of civilized society through the "other"—for example, the "noble savage" of Jean-Jacques Rousseau (1712–78) or the sexually liberated Samoans of Margaret Mead (1901–78). Although third-world natives were not noble, nor Samoans free of sexual anxieties, claiming otherwise embarrassed the "civilized." In the age of pluralism, extraterrestrials often served as a counterpoint to shame humanity, and they have continued to play this role in modern science fiction and occasionally in SETI itself.

Pluralism in the Nineteenth Century

With many scholars believing in extraterrestrials, the public's credulity is not surprising. When the *New York Sun* claimed in 1835 that English astronomer Sir John Herschel (1792–1871) had seen moon people with his new telescope, many readers believed the hoax, showing how deeply the idea of extraterrestrial life had penetrated the popular consciousness. Not for the first time would fantastic stories of contact take in the public.

Several scientists suggested signaling extraterrestrials of our existence. In the 1820s, the mathematician Karl Friedrich Gauss (1777–1855) proposed clear-cutting large parts of the Siberian pine forest to illustrate the Pythagorean theorem, thereby demonstrating to the moon's inhabitants that Earthlings were intelligent. Joseph Johann von Littrow (1781–1840), director of the Vienna Observatory, also proposed a system of mathematical signaling. He suggested digging huge geometric trenches in the Sahara Desert, filling them with kerosene, and setting them alight, in the hope that other races in the solar system would notice. The French physicist Charles Cros (1842–88) proposed a system of mirrors to reflect sunlight toward Mars.

By the mid-nineteenth century, metaphysical pluralism was facing sharp criticism. William Whewell (1794–1866), in his *Of the Plurality of Worlds* (1853), examined the argument from analogy and found no hard evidence and concluded that humanity is unique. In a way, Whewell foreshadowed today's SETI deniers who believe humanity unique in the universe. John Stuart Mill (1806–73) also objected to the argument from analogy, noting that whenever two phenomena have more dissimilarities than similarities, the unknown aspects are less likely similar. His example was alleged moon life. In Mills's opinion, without more astronomical knowledge, analogical reasoning was invalid for pluralism.[7]

In the second half of the nineteenth century, pluralism had the good fortune of three great scientific advances: the nebular hypothesis, spectrum analysis, and the theory of biological evolution. The nebular hypothesis was a new view of the solar system's origins.[8] According to this hypothesis, the solar system was formed when a huge cloud of gas and dust, called a nebula, collapsed and was pulled in by gravitational forces. The nebular hypothesis boosted pluralism because it implied that the same process formed other stars and planets.[9]

Spectrum analysis began with the discovery that the spectrum (or light frequency) of each chemical element produces an emission line with a unique pattern. Using this method, scientists soon discerned Earth's elements on the solar system's planets and moons. Spectrum analysis also revealed that stars contained the same elements as the sun. In other words, stars, no matter how distant from our solar system, even trillions of light-years, are simply other suns. Like our sun, they too may be surrounded by orbiting planets (since the 1990s a proven fact).

All that remained was to explain extraterrestrial life without referring to a deity. The theory of biological evolution did this. Proposed by Charles Darwin (1809–82) and Alfred Wallace (1823–1913), evolutionary theory made the development of terrestrial life a natural process. Survival of the fittest, or fortunate, was the key; it forced species to adapt to changing environments or perish. According to Darwin, a species surviving over the millennia, if not millions of years, might change physically and eventually become a new species.[10]

Darwin's evolutionary theory had a staggering implication for pluralism. If life arose on Earth and developed from lower to higher forms, the same natural process could take place elsewhere, especially since the universe contained the same elements throughout. Evolution now replaced teleology in the pluralist schema. No longer were extraterrestrials necessary to give other worlds a purpose. As on Earth, local conditions determined life's emergence.

The Strange Career of Percival Lowell

In 1877, the Italian astronomer Giovanni Schiaparelli (1835–1910) announced that the Martian surface had straight features, which he called "canali." Unfortunately, this Italian word, which means "channels," was mistranslated as "canals," feeding the imagination of a generation prepared by pluralistic thought and recent science to believe in extraterrestrial life.[11]

The best-known canal chauvinist was Percival Lowell (1855–1916), who claimed that his telescope revealed canals on Mars. Scientific in approach and the author of scholarly papers, Lowell could not be dismissed as a quack. Although knowledgeable, Lowell was wrong nonetheless. His career shows how agendas can twist the faculties of even the intelligent and honest.

Lowell was an old-money, proper Bostonian who could have quietly enjoyed life and its privileges. Instead, Lowell was fearful and alienated. He worried over the relative decline of Boston and its elite, eclipsed by the crass new money of the industrial and oil barons who preferred to live in New York City. Lowell also worried over dangers from below. The lower classes were increasingly assertive and, along with the ever-increasing "immigrant horde," threatened to swamp the WASP elite. Proper men were not even safe from the distaff side. The emerging feminist movement threatened the sexual hierarchy he took for granted.[12] Lowell convinced himself that his Martian observations validated his social and political views.

Like his Puritan forebears who believed themselves chosen by God,

Lowell had the itch to preach. Lowell needed a mission and he discovered Mars. He concocted a mythology about the Red Planet, and if his theories had held up he might have become as famous as Galileo. Instead, he was a shooting star, temporarily the object of much attention but destined to fall, recalled only by astronomers, who see his Martian obsessions as somewhat of an embarrassment.

Lowell did his research in Flagstaff, Arizona, where he built his own observatory on a hill ("Mars Hill" he called it). Lowell bragged of the great viewing in the fog-free, dry climate. He always insisted that superior telescopic investigation lay behind his Martian pronouncements. Lowell never realized that he was guilty of pareidolia, the imposition of known images on random patterns. In later times, stone "faces" would be seen on Mars, as if Martians had created their version of the Great Sphinx. In Lowell's day, canals dominated the news. The Suez Canal had been completed in 1869, and its French builders were digging another canal in Panama. Hence, Lowell saw canals on the Martian surface.[13]

Lowell's Mars was dying. Theory was responsible. Lowell subscribed to a deterministic evolution in which the cycle of birth, growth, and decay was universal. Since Mars was older than Earth, Lowell believed the Red Planet was suffering from desiccation, on Earth beginning, but on Mars well advanced with most of its surface already turned to desert. Lowell's Martians were desperately building canals to tap water in the polar ice caps, a doomed attempt to fend off extinction.[14]

Since Martian life predated Earth's, Lowell believed Martians were further evolved than humans. Martian superiority as well as the terminal decline of the Red Planet convinced Lowell he could deduce the essentials of Martian civilization. Like the pluralists of previous generations, he used alien behavior as a counterpoint so he could pontificate about humanity's alleged defects. Unlike Earthlings, beset with divisions and hatreds, Lowell's Martians had united for the common good and had abolished war in order to concentrate on their canals. Lowell's Martians were cooperative, but too smart for democracy, which Lowell equated with confusion. The best and the brightest ruled on Mars, with the lower classes, conscious of the greater needs of society, allowing the intellectual elite to rule. Lowell let this be known. In 1911 he told Arizona miners, a radical lot, that labor-capital divisions were absent

on Mars, whereas "ignorance and ambition dominate terrestrial politics." Lowell had persuaded himself that the equivalent of the Boston
Brahmin elite ruled Mars.[15]

Lowell wrote and spoke well. His witty lectures and lucid publications made him a celebrity, although he was somewhat of a killjoy. His
picture of advanced Martians disturbed much of the public, which preferred a superior humanity. Lowell did not yield. He professed a cold
detachment, reporting Martian achievements and travails as if he were
describing a beehive. He created a frenzy in which Mars took on a life
of its own in the popular mind.[16]

Professional astronomers could not ignore Lowell. The more they
examined his canals, the greater their skepticism. When they looked
at the "lines" on the Martian surface, they did not see canals; they saw
lines natural in origin. Some astronomers blamed Lowell's twenty-
four-inch refractor, a smallish telescope, for misleading him. George
Ellery Hale (1868–1938), the founder of the Mount Wilson Observatory, which housed a sixty-inch reflector, could not see the canals. W.
W. Campbell (1862–1938) used a different approach; his spectrographic
analysis could not detect water vapor on Mars.

In 1909, astronomers had an opportunity for a very close look at the
Red Planet because Earth and Mars were in opposition, that is, their
orbits had drawn them unusually close. Here was a chance to settle the
controversy. Lowell's opponents saw no evidence of canals, and as far
as they were concerned the Martian question had been resolved. After
1909, with Mars and Earth drawing farther apart, no new evidence was
possible.[17]

Lowell himself paid more attention to other astronomical matters,
such as his suspicion that another planet lay beyond Neptune. After his
death in 1916, the discovery of Pluto in 1930 was a belated vindication,
but only for the rest of the twentieth century. In 2006, the International
Astronomical Union downgraded Pluto, defining it as a dwarf planet.

The verdict on Lowell? He raised the public's awareness of extraterrestrials, and he greatly affected science fiction.[18] Perhaps his greatest
contribution was seeing the fate of the older Martian civilization as predictive for humanity. This intuitive belief persists today, though with
extraterrestrials replacing Martians, and it is a justification for the SETI

enterprise. In the short run, though, Lowell lost not only the battle but also the war. For years afterward, astronomers would be very careful, if not derisive, whenever alien life was discussed. They feared being seen as following in Lowell's ludicrous path. In science, appearing ridiculous is a greater offense than being wrong.

Astronomers also had solid theoretical reasons for ignoring the possibility of extraterrestrials. They were suspecting that the solar system, with the exception of Earth, was lifeless, at the very least lacking intelligent life. Besides, a new view of the solar system's formation had emerged. In 1917, the British astronomer James Jeans (1877–1946) proposed the encounter theory in lieu of the nebular hypothesis. Jeans held that a star wandered very near the sun, and although it did not collide with the sun, the star's gravity sucked out some of the sun's gas, which became a feedstock for the formation of the planets. Since encounters like this are rare in the vast expanses of space, the encounter theory implied the rarity of extrasolar planets. The great pluralist tradition had ended, or so it seemed.[19]

2

Science Fiction and Ufology

Strange Cousins

Percival Lowell's obsession with Mars had the toxic effect of making astronomers reluctant to discuss extraterrestrials. Besides, James Jeans's encounter theory made them unlikely. Banished from astronomy for at least a generation, extraterrestrials needed a new home, and they found it in science fiction. Only here, with Earth thoroughly explored, could the reading public find fantastic tales of strange beings and places. Although Lowell was dead, his Martians, transformed into gaseous monsters, space shifters, and wise beings, lived on.

Legions of science-fiction writers invented strange, quirky extraterrestrials who beguile, frighten, or instruct. The stranger the beings, the more they capture readers' attention. Not even Hieronymus Bosch, the Renaissance artist of the bizarre, imagined such odd visions of life. Through the literary genre of science fiction, extraterrestrials have intruded and settled into the human consciousness.

Science Fiction and Its Agenda

The bizarre creations of science fiction often have an ulterior motive. In the pluralist era, extraterrestrials were used to comment on human behavior. Science fiction has done the same. It takes present trends and extrapolates from them, offering possible outcomes, an imaginable tomorrow. The peek into an exotic future hooks readers, but whether they realize it or not, they are being regaled with commentaries on the present.

Although much science fiction has been space opera, exotic tales of strange people and creatures that could no longer be set in the world's remote parts, it is also a medium for spreading ideas. Unlike scientific thought and philosophy, one expressed in the language of mathematics and the other bundled in abstractions, science fiction personalizes ideas, performing the same function that a medieval morality play did for the profundities of theology. By making the abstruse easily understandable, science fiction, although a commercial venture, served to educate the masses.

Perhaps no writer of science fiction preached better than H. G. Wells (1866–1946), one of the genre's founders. His *The War of the Worlds* (1898) was a best-seller in its day, is still read, and has inspired imitations as well as two Hollywood movies. Wells's secret was frightening the public. He portrayed Martians as technologically superior invaders bent on exterminating humanity. Landing near London, the Martians march to the British capital, then the political center of the world, easily overcoming all opposition.

Percival Lowell had prepared the public. He had created a desiccated Mars whose inhabitants strove to delay the inevitable. Wells borrowed Lowell's scenario of a doomed Mars and made one crucial change: his Martians were unwilling to fade graciously into extinction. They were like the many Earthlings who seek better lives by exploiting others. To make sure his readers got the point, Wells laid it out clearly by noting that his fellow Britons had built their empire by subjugating and occasionally exterminating the weak, such as the natives of Tasmania. They had no right to complain about the Martians.[1]

In the Martian novels of Edgar Rice Burroughs (1875–1950), the Martians stay home. The traveler is the Virginia-born hero John Carter, who befriends the good Martians, fights the bad ones, and ends up marrying a Martian maiden. Burroughs's Martian tales sold well. John Carter is brave and noble, a hero who reinforced early twentieth-century optimism.[2]

Burroughs wrote to make a living, but as often happens with science-fiction authors, he was not above spreading a message. In *Princess of Mars*, Burroughs not too subtly criticizes socialism. The Martian Green Men own everything in common, socialist style. They are the villains.

Although Burroughs's Mars contained evil Martians, none threatened Earth. This vision of safe extraterrestrials may have influenced the young Carl Sagan, who, like many boys, read and enjoyed the John Carter novels. Sagan grew up to be the best-known enthusiast for pluralism in the late twentieth century. His pet name for Mars was "Barsoom," which was what Burroughs's Martians called their planet.[3]

Burroughs's Martian adventures were entertaining, but they were no match for Wells's dark vision. The public has usually been more receptive to bad news. Wells had struck a nerve by updating the Apocalypse of John the Evangelist. To the traditional four horsemen of death, famine, war, and pestilence, Wells added a fifth: the genocidal extraterrestrial.

Even the usually sedate *New York Times* seemed fearful of extraterrestrials. In 1919, when Guglielmo Marconi, inventor of the wireless radio, proposed using radio to contact extraterrestrials, the *New York Times* sternly warned, "Let the Stars Alone." Its editorial insisted that humanity was not ready for superior beings.[4] The newspaper might have been worrying that extraterrestrials with superior IQs. would embarrass Earth's intellectuals; or, schooled in the dark tales of science fiction, it had more down-to-earth fears: the Martians of H. G. Wells.

In 1938, Orson Welles showed how well a fraction of the public had internalized Wells's Martians. Welles starred in radio's *Mercury Theatre* and, on October 30, along with other actors he performed a free adaptation of *The War of the Worlds*. Instead of England, the Martians landed in New Jersey, destroying all opposition and marching to New York City. Although the extent of the public's panic was later exaggerated, itself becoming an urban legend of sorts, some listeners believed that an invasion was actually taking place and became very nervous. The panic was later blamed on the uncertainties of the Depression era and the more immediate international situation. Hitler had been in power for five years and had been threatening war in the summer and fall of 1938 unless Germany could annex the Sudetenland, a portion of Czechoslovakia. The public had gotten used to radio news bulletins announcing the latest European crisis. A sudden news bulletin of a Martian invasion was more bad news—only worse.[5]

Besides commenting on current events, science fiction borders on religion when it searches for something beyond ourselves, dealing with ultimate questions such as human destiny and the purpose of life.[6] The science-fiction writer famous for the destiny question was the Englishman Olaf Stapledon (1886–1950). In his *Last and First Men: A Story of the Near and Far Future* (1930) and *Star Maker* (1937), Stapledon, an atheist, proclaimed evolutionary biology as the engine driving destiny. Evolution to intelligent beings was a regularly occurring event, not a onetime accident. *Last and First Men* spans two billion years during which eighteen species of humans evolve, first on Earth, then on Venus, and finally on Neptune. The eighteenth is the last species and it achieves a spiritual status. Near the end of the novel, Stapledon says that at least twenty thousand worlds in the galaxy have produced life, many of them belonging to the Interstellar League.

In *Star Maker*, Stapledon develops the theme of animate worlds. The narrator roams the universe, sees countless beings, even God himself, who is indifferent to his creation—no Christian god of love here—but it does not matter. Humanity has evolved to the point where it no longer needs God. Although Stapledon presents the future, he drops a subtle hint about the present: humans should accept space travel and contact with intelligent extraterrestrials as part of their destiny.[7]

Human destiny is also central in *Childhood's End* (1953) by Arthur C. Clarke (1917–2008). The Overlords come to Earth and create a utopia, which bans war and cruelty. The Overlords use a gentle approach, ending bullfighting, for example, by having spectators feel the bull's pain. It turns out that the Overlords are preparing humanity for a sudden evolution into a superior mental entity and for absorption into a cosmic Overmind.[8] Clarke's teleology is somewhat similar to the theological vision of the French Jesuit Pierre Teilhard de Chardin (1881–1955), who wrote of humanity expanding in consciousness and transcending itself. Whether Teilhard was straying into science fiction is beside the point. The Vatican discerned religious deviance and silenced Teilhard.

When scientists speculate on extraterrestrial life, they often betray a debt to science fiction. The SETI belief in benevolent aliens who help humanity echoes more than one science-fiction piece. At the same time Wells invented fiendish Martians, Germany's Kurd Lasswitz (1848–1910) presented ethically evolved Martians who tried to uplift benighted Earthlings.[9] His novel *On Two Planets* was little read in the English-speaking world, but it was quite popular on the European continent. It was a favorite of rocket pioneer Wernher von Braun (1912–77). Not accidentally, when Morrison and Cocconi published their seminal SETI paper, they denied writing science fiction. They were aware of the porous boundary separating scientific speculation on extraterrestrials and the world of literature.

SETI advocates like to believe that the galaxy contains long-lived, advanced civilizations. By coincidence or not, this utopianism parallels a theme in science fiction, exemplified by the *Star Trek* television series and its sequels, which ran from the 1960s into the new millennium. Along the way, *Star Trek* became the most popular science-fiction saga in television history. Pretending to be space opera, *Star Trek* episodes were often morality tales with a progressive outlook. In *Star Trek*'s universe, humanity survives its growing pains, renounces war, and joins the United Federation of Planets, whose only substantive worries are outsiders, such as the Romulans, whose militarism echoes Earth's wicked past.

Ufology

Although in the second half of the twentieth century, science fiction continued to preach—science-fiction writer Ray Bradbury (1920–2012) and the *Star Trek* saga come to mind—the preaching mode developed a new literature, the participatory theater of ufology, in which average people claim to have seen or spoken to extraterrestrials and, in some cases, to have been violated by them. Unlike Lowell's Martians, safely distant by millions of miles, extraterrestrials now walked among us.

Ufology has been the great popular myth of the second half of the twentieth century. Its devotees or "ufologists" believe that extraterres-

trials are responsible for the flying-saucers phenomenon. Many ufologists go further and claim that extraterrestrials have contacted Earthlings, that they interfere in human affairs, and that a vast conspiracy
involving government and private institutions hide all this from the
public.

Ufology was born on June 24, 1947, when Idaho businessman Ken-
neth Arnold flew his small aircraft near Mount Rainier in Washington
State and saw something he could not identify. He later recalled flying
objects skipping like saucers over water. A reporter simplified Arnold's
words and wrote that Arnold had seen "flying saucers." Other observers
also claimed to have seen mysterious objects in the sky, and the term
"flying saucer" caught on.

The U.S. Air Force dismissed Arnold's flying objects as natural phenomena, as it did with future sightings. The public, goaded by the press,
was suspicious. Rather than believing flying saucers were unexplained
tricks of nature, the skeptics suspected that the unidentified flying objects, or UFOs, were secret military vehicles or perhaps Russian aircraft, or even of extraterrestrial origin. They were sure of this much:
the U.S. government was not being candid.[10]

Eventually, a human being claimed to have spoken to an extraterrestrial. George Adamski (1891–1965) was that person. An ice cream
salesman from Mount Palomar, California, Adamski defined himself as
a "philosopher, student, teacher, saucer researcher." On November 20,
1952, Adamski announced that a flying saucer had landed in the California desert and that he "made personal contact with a man from another world." A handsome, blond alien said that extraterrestrials were
upset over humanity's atomic weapons. The previous year, a Hollywood
movie, *The Day the Earth Stood Still*, featured an alien who sternly lectured humanity on its wicked atomic bombs. Adamski's alien was less
censorious, though, suggesting that he would use natural forces instead
of violence to curb human aggression.[11]

Adamski is easily dismissed as a harmless hoaxer, especially since
the photos he submitted as proof appeared crudely faked. A closer look
at his tale reveals a serious side. He belonged to the tradition of using extraterrestrials to soften and disguise a message of dissent. Adamski came forward at the height of the McCarthy era, when many

Americans felt intimidated. By having extraterrestrials decry atomic weapons, Adamski was avoiding identification with the political left and staying clear of McCarthyite censors.

Adamski achieved a middling fame. He published books, appeared on television, spoke abroad, and was granted an audience by Queen Juliana of the Netherlands. This success would be noticed. Other contactees emerged and told mostly the same story of aliens with a message.[12] Contactees were akin to the historical people who transcended their ordinary lives by claiming a cautionary or uplifting message from a superior intelligence. The public could either believe or reject.[13]

Within a decade, contactees had become boring. Like Hollywood gore movies that quickly jade the public, forcing moviemakers to raise the shock level with more blood and butchery, ufology needed to expand its narrative. The direction of that expansion came when Barney and Betty Hill launched the abductee era in 1971. The Hills' tale was simple. While driving their car one night, they were kidnapped by aliens and medically examined. As with Adamski, the Hills started a trend. New abductees told of violation and rape, even of insemination, pregnancies, and forced abortions.[14]

Ufology has intruded into our modern age. It is a new medium of narration whose alleged authenticity makes it more compelling than traditional science fiction, in which the real world disappears only while the text absorbs the reader or the movie envelops the viewer. Whereas life reverts to normal after watching *Star Trek* or reading Ray Bradbury, the abductee's tale evokes permanent anxiety, since we too might meet an alien, and the abductee's fate may be ours. Ufology has broken the barrier between a stranger's misfortune and us. We no longer empathize; we fear. Ufology includes all of humanity in the drama; it is grassroots folk theater writ large. By acting out science fiction, ufology has become a relative of science fiction, whether a close or distant cousin is beside the point.

Ufology is also a political counterculture, though not of the 1960s variety, which centered on personal freedom. Ufology is political in that it distrusts experts; it rejects the rationalism that permeates what passes for the establishment. Ufology is an outlier, finding no succor in the mainline churches, the leading universities, the scientific and medical authorities, and the allies of all these in the respectable news media. Journals of political opinion such as *The Nation* and *National Review* disagree on much, but they agree that extraterrestrials are not lurking among us or that a huge conspiracy exists to prevent the public from knowing about it.[15]

If extraterrestrials often show themselves to humans, the implications are staggering. Extraterrestrials avoid the glitterati. They have no time for leading politicians, bankers, and clerics; they instead reveal themselves to the little people. Instead of approaching the Harvard University faculty, extraterrestrials have sought Bubba in rural Texas, suggesting that they are quintessentially democratic and anti-elitist. Another view is possible. Extraterrestrials are so advanced that they regard semi-literates as just as interesting, if not more, as Nobel Prize winners.

Various explanations have been given for ufology. Ufology has been dismissed as Cold War stress, as the dumbing down of America, as a species of New Age religion that parodies Christianity with its own pseudo angels and demons. Ufology has also been seen as the creation of irresponsible media that feed the narrative through sensationalist books, articles, and TV shows. They disseminate alien images and stories of sightings and encounters throughout the popular culture. Not surprisingly, according to some polls, many Americans believe that extraterrestrials are visiting Earth.[16]

SETI and Ufology

The SETI community is skeptical of ufology, finding it preposterous that aliens travel hundreds if not thousands of light-years through space, a voyage that testifies to technology beyond human comprehension, merely to give unlucky humans an anal probe. Defenders of the abductees note the sincerity of the "victims" and go on the offensive, accusing the SETI community and the rest of the scientific establishment of closed-mindedness, of refusing to recognize a reality beyond their instruments and mathematical models.[17] Yet ufology has performed a service for SETI.[18] It has raised the average person's awareness of the possibility of extraterrestrial life, giving SETI a ready acceptance with the public. On the other hand, because of ufology, SETI astronomers in their early years needed to convince the scientific establishment of their legitimacy.

The SETI community (with an exception or two) prefers to ignore ufology, a strategy that can backfire, for the public, including the occasional politician, have confused the two. When NASA got involved in SETI, several politicians were derisive, labeling SETI an exercise in ufology. In 1993 Congress voted to end NASA's SETI program. SETI survived by privatizing, relying on donations from sympathetic millionaires.

SETI has had the misfortune of being sited in a postmodern age, in which truth is uncertain, authority is suspect, and all views, like ufology, possess some validity. A television documentary can give equal weight to orthodox scientists such as SETI pioneer Frank Drake and to committed ufologists. An astronomer who lectures on the possibility of extraterrestrial life may be confronted by an audience member who cites contactees and abductees as evidence. The astronomer has the years of study, the advanced degrees, the research, the peer-reviewed publications, but in the ufologist world these orthodox credentials count for naught. In our postmodern age in which anyone can be an expert, ufologists demand respect.

Ufologists and SETI people share a belief in advanced extraterrestrials. They disagree, however, on the time table of contact. Ufologists insist that contact has already taken place, whereas the SETI community speaks of contact in the future.[19] Both also subscribe to a culture of dissent. When ufology insists that the U.S. government knows of extraterrestrials and conceals their malevolent acts, it assumes, as do most conspiracy theories, that government lies much of the time, if not all the time.[20]

The SETI community also has a dissenting agenda: a dissent born of despair, the fear of many scientists that humanity may commit suicide through overpopulation, pollution, or atomic warfare. During the Cold War, when the possibility loomed of Russian and U.S. missiles destroying the planet, extraterrestrials were seen as ethically superior to a selfish humanity. It was assumed that advanced beings had survived lethal technologies and might be gracious enough to share their survival strategy—in short, that there had been an ethical evolution as well as a biological evolution in the worlds of extraterrestrials. Contact thus became a means of salvation.[21] All this echoed the message of Adamski and other contactees, something that the SETI community ignores.

3

The Road to SETI

Fearing the Future

Why was extraterrestrial contact important to SETI activists? One reason was their pessimism, their despair over humanity's future and their hope that extraterrestrials might have a solution for humanity's ills. SETI activists believed that an advanced extraterrestrial society that had survived its version of self-induced extinction might give useful advice.

The pessimism of SETI activists typified much twentieth-century thinking. Intellectuals were often gloomy. They had lost faith in the old verities, having dismissed traditional truths of religion and philosophy as culturally inspired and worthless, relativism having replaced the universality of ethical standards. The comforts of God and culture were no more.

Western intellectuals had never been this gloomy. During the Enlightenment, the name given to the intellectual movement of the eighteenth century, they confidently believed in progress; they saw knowledge increasing over time and behavior improving. This optimism persisted into the nineteenth century, and for good reason. Advances in medicine, technology, and science had been astounding. By 1900, more people than ever lived in the Western world, yet they were living longer and were better fed than their ancestors.

The Twentieth Century: Age of Pessimism

Then like a bolt of lightning came the world war of 1914–18. World War I showed that all those wonderful machines and gadgets of the industrial age that improved life could also be perverted to kill and maim. The ingenuity that had produced the railroad and the telephone also produced rapid-fire guns, tanks, fighter planes, and poison gas. Millions of European soldiers died over what today seems to be piddling political gain. European society never regained the certainty and swagger of the nineteenth century.

For Europe, World War I was not only a social catastrophe but also an economic disaster. The United States had entered the war late—in 1917—and suffered far fewer casualties, and yet it became the real victor. Although Great Britain and France were also victorious over Germany, they had consumed their wealth to fight the war. The United States, a debtor nation to Europe prior to the war, had lent billions to its allies and emerged from the war as a creditor nation. To its industrial might the United States added the dominance of world finance. No wonder that gloom deepened on the Continent, especially among intellectuals. The wild success of Oswald Spengler's *The Decline of the West*, which foresaw the end of Western civilization, was no accident.

By contrast, Americans felt good in the 1920s, more worried by Prohibition than by the state of the world. Suddenly, in 1929, the stock market crashed and set off the Great Depression. Instead of rising living standards being the norm, as in the 1920s, the United States of the 1930s saw mass unemployment and impoverishment. Not only did the diminished paycheck, if there was one at all, become the lot of the average American, there was also the unsettling feeling that the whole American experiment was at fault.

An alternative was Communism, the movement of the 1930s that both frightened and fascinated Americans. Communism was established in Russia as a result of the Revolution of 1917. By the 1930s, although the brutal dictator Joseph Stalin (1878–1953) had become the face of Communism, the ideals of Communism remained seductive. Its secret was that it fed on pessimism and offered optimism. It embodied pessimism, in that its devotees had lost all confidence in old ways and traditions, but it also offered a utopian society, an aspiration since antiquity. It had to be understood, said Communists, that the root cause of war, crime, and poverty was the unequal distribution of wealth and that corrective steps had been already taken in the Soviet Union, which had liquidated the capitalist and landowner classes. Communism, Russian-style, had to be accepted in order to improve the world.

Nearly all Americans rejected Communism, but many intellectuals flirted with the new gospel from Moscow, some going so far as to join the Communist Party. A Communist turn signified rebellion, and since the young often like to shock, some college students joined campus Communist clubs. Wearing a red tie to class and giving the Communist salute annoyed college administrators far more than a panty raid.

Politics and the Atomic Bomb

One of those college students attracted to Communism was Philip Morrison. Never afraid of controversy, Morrison had joined the Communist Party during his days at Carnegie Tech. In 1939, while a graduate student at UC Berkeley, Morrison maintained the membership, a connection that may explain why he could later say that he knew more Americans who died fighting the fascists in the Spanish Civil War (1936–39) than who died in World War II.[1]

The Great Depression that made some embrace Communism ended during World War II. Although the United States did not enter the war until the Japanese attacked Pearl Harbor in December 1941, the nation had already begun to rearm, thus creating jobs in defense plants. With the entry into the war, the need for munitions and the draft sucked up surplus labor from the ranks of the unemployed. The Great Depression finally ended.

Although prosperity had returned, new concerns replaced old ones. In 1939, Albert Einstein had written a letter to President Roosevelt, warning that Nazi Germany might be building an atomic bomb. The result was the Manhattan Project, the supersecret crash program to beat the Germans in the atomic bomb race.

Morrison worked on the Manhattan Project. Although of draft age, he did not serve in the armed forces. He was disabled, struck with polio at the age of four and having to wear leg braces. Besides, he was too useful to the war effort as a scientist. At UC Berkeley, Morrison had pursued a PhD in physics, studying under J. Robert Oppenheimer (1904–67), who was to direct the Manhattan Project. Morrison first worked on the atomic bomb at the University of Chicago and in 1944 went to the center of bomb research at Los Alamos in New Mexico. He also helped to transport the finished atomic bombs to the Pacific. On the island of Trinian he led the team that assembled the bomb that was dropped on Hiroshima in August 1945.

The Atomic Scientists Movement

In June 1945 the atomic bomb had been ready, too late for the war in Europe, with Germany having already surrendered in May, but still available to drop on Japan. Since Japan did not have an atomic bomb, many atomic scientists had second thoughts about using their dooms-day weapon.

University of Chicago professor James Franck (1882–1964), Nobel laureate in physics in 1925, who had been deeply involved in the Manhattan Project, penned a famous report to the secretary of war on June 11, 1945. Speaking for many atomic scientists, Franck saw the atomic bomb in political and moral terms. He called for demonstrating the bomb's power to Japan, and only if Japan refused to surrender did he consent to "perhaps" using the bomb, provided the United Nations and domestic public opinion approved.[2]

Franck's plea never had a chance. Merely demonstrating the bomb would allow the Pacific War to continue while the Japanese mulled their options. Meanwhile, U.S. soldiers would continue to die, young men far removed from the rarified world of atomic physics and mostly drawn from the working class. The atomic scientists did not know the common man, but President Harry S. Truman not only knew the common man, he was the common man: the last U.S. president to have skipped college. Whereas Truman was prey to the tribalism of American society, Franck and the atomic scientists had gone beyond it, reborn citizens of the world. Franck himself was German, a refugee from the Nazi racial laws, who had lived in Denmark before coming to the

United States. In the calculus of Franck and other atomic scientists, the death of a Japanese citizen carried the same weight as an American's death. President Truman felt otherwise. When the atomic bombs were ready, he authorized dropping them, first on Hiroshima on August 6, and three days later on Nagasaki. To the day he died, Truman insisted he never regretted having saved American lives.

The atomic bombs did quickly end the war, but as Franck had stated, in the long run the doomsday weapons were primarily a political challenge. With the end of World War II a new war began, the so-called Cold War between the United States and the Soviet Union, a "war" in which neither side fired at each other, although the danger always loomed of a shooting war involving atomic weapons.

Since atomic scientists were basking in the prestige of the new atomic age and everyone knew they were smart, they believed they could educate the public about nuclear war and its dangers. In late 1945 a group calling itself the Atomic Scientists of Chicago started publishing the *Bulletin of the Atomic Scientists*. Every issue of the *Bulletin* contained a doomsday clock in which the closer the minute hand crept to midnight, the closer the danger of Armageddon. In addition to warning the public, the atomic scientists had a plan to prevent nuclear war: creating a world government to control nuclear energy. Oppenheimer spoke for many atomic scientists when in April 1947 he helped found the United World Federalists, whose goal was ending national sovereignty. The need for action seemed even greater when in August 1949, far sooner than U.S. intelligence had expected, the Soviet Union exploded an atomic bomb.[3]

Although the Soviet bomb made nuclear war seem closer, the atomic scientist movement was running out of steam. It embodied the usual defect of academic activism: the supposition that enormous expertise in a single field leads to intuitive insights on all sorts of complicated societal and political issues. World government was an ill-conceived solution; it presupposed nations so trusted each other they could renounce their national sovereignty. If this level of trust existed, world government was unnecessary. Besides, frightening the public into making politicians act was unrealistic; it was a scenario from science fiction. In H. G. Wells's *The World Set Free* (1914), humanity reacts to a

French nuclear attack on Berlin by bypassing governments and creating a supranational organization to impose peace. In the real world of the late 1940s and the 1950s, Americans saw the Communist bomb as greater reason to back their government and distrust the Soviet Union. The atomic scientists themselves lost credibility when many of them, most notably Oppenheimer himself, were accused of having Communist ties.[4]

Activist atomic scientists and their supporters had a rough time in the 1950s. In November 1952 the United States tested a hydrogen bomb, a weapon a thousand times more powerful than an atomic bomb, only to have the Soviet Union in less than a year respond with its own H-bomb explosion. The doomsday clock of the atomic scientists crept to two minutes from midnight. To worsen matters, even if the United States and the Soviet Union avoided war, the very testing of nuclear weapons was deadly, giving off radiation that caused birth defects and cancer. Although radioactive fallout spurred a movement against nuclear testing, many Americans of the 1950s were not alarmed; they were moving to suburbs, buying big-finned cars, and more worried by the popularity of Elvis Presley and rock music.[5] In December 1959, Eugene Rabinowitch, editor of the *Bulletin of the Atomic Scientists*, complained that Americans little appreciated science and abdicated many key decisions to politicians and the military.[6]

Philip Morrison

If only humanity could find a new grounding for society, save the world, and advance science at the same time—achieving this trifecta was Philip Morrison's dream. Already radicalized during the Great Depression, Morrison grew gloomier with the coming of the atomic age. Like many other Manhattan Project scientists, Morrison would spend the rest of his life worrying about the future and opposing nuclear weapons.

In 1946 the Federation of American Scientists published *One World or None*, an anthology in which scientists discussed the atomic bomb's danger and the need for federal world government. Morrison contributed a gripping chapter, "If the Bomb Gets Out of Hand," about his visit to Hiroshima a month after the bomb had been dropped. Having measured the damage and the radioactivity, he compared the Hiroshima devastation to a battlefield where a conventional bomb had directly hit every building. He noted the quiet reproach of the Japanese scientist he had interviewed. More than half of Morrison's essay described what would happen if a Hiroshima-type bomb detonated just above the corner of Third Avenue and East Twentieth Street in New York City. Morrison estimated that 300,000 New Yorkers would die. His conclusion: "If the bomb gets out of hand, if we do not learn to live together so that science will be our help and not our hurt, there is only one sure future. The cities of men of Earth will perish."[7]

In 1946 Morrison accepted a position at Cornell University, where his interest in students and his splendid lectures made him excel. Morrison had that rare gift of being articulate and of explaining science in plain English. Although a respected scientist, with a reputation for thinking differently, in his pre-SETI days Morrison seemed to get more attention for his politics.

Although he had joined the Communist Party during his college days, at this time Morrison was no longer a Communist, although still a progressive. The flirtation with Communism revealed much about Morrison's mind-set. Embracing Communism had been a last resort of sorts, a belief that politics and culture were hopelessly outdated, if not downright dysfunctional and evil. Embracing Communism signaled a willingness to take a very bold step to effect radical and beneficial change.

In 1951 Morrison was a sponsor of the American Peace Crusade, which the U.S. government labeled a Communist front. Morrison rationalized that an effective American peace movement had to include all Americans, even partisans for the Soviet Union. Two years later, when testifying before the Senate's Internal Security Subcommittee, he admitted to his Communist past but refused to name his former associates in the Communist Party. Morrison's politics led to trouble at Cornell, where sporadic attempts were made to have him fired. In 1965 he moved to MIT.[8]

Morrison always remained a progressive. In 1998, seven years before his death, he published *Reason Enough to Hope*, in which he urged the United States to reduce its military arsenal and rely on its allies for support. Partial disarmament was not an end here but a means. A militarily weakened United States would need to cooperate with its allies, creating world federalism through the back door.[9] Morrison was returning to the atomic scientist dream of the 1940s.

The Population Explosion

Along with fear of nuclear annihilation that began at the end of World War II and remains to the present, another seeming threat to the future emerged in the postwar years: overpopulation. World population had been increasing for centuries. During the twentieth century, its rate of increase escalated. Many were unnerved.

Demographic pessimism was nothing new. It dated to Thomas Malthus (1766–1834), whose *Essay on the Principle of Population* sounded the modern alarm. In 1798, Malthus had predicted that population in the nineteenth century would geometrically increase while food production lagged far behind, causing starvation and riots by century's end. A new barbarism would emerge in which people stalked and killed each other for the last scraps of food. This was a hard view of the future, and a fallacious one, for never has the passage of time proven a scholar so wrong. Malthus totally missed the significance of the Industrial Revolution, already begun in the England of his day. Factories and farmers replaced workers with machines; labor inputs no longer needed doubling to double output. Although population did skyrocket by the end of the nineteenth century, so did production of food and consumer goods.

Malthus managed the unusual feat of being both wrong and influential. His failed predictions became the basis for a neo-Malthusian model, which held his original pessimism essentially correct, only wrong in the timetable. According to this model, the Industrial Revolution had only delayed the inevitable. If population continued to increase, the human race would find itself swollen beyond hope, consuming limited, ever-shrinking resources. A social cataclysm was inevitable.[10]

For neo-Malthusians, the twentieth century was alarming. In 1900, Earth's population was roughly 1.5 billion. In the ensuing decades, millions were slaughtered in two world wars, yet world population kept growing. By 1950, that population had reached an alarming (to neo-Malthusians at least) 2.4 billion.[11] By 1960, alarmists had more reason to fret; the postwar baby boom had pushed the population beyond 3 billion.[12] Inasmuch as Americans had a high standard of living, they should have shown restraint. Instead, they indulged themselves. Their fertility rate in the 1950s reached a high of 3.8 children for each woman, although a rate of 2.1 would have been sufficient to replace the existing population.[13] The baby boom gave no signs of abating and led many to ask when living standards would decline. Malthus was being proven correct, or so it seemed. Like locusts, humans were overbreeding and were fated to devour all food resources, and, like locusts, they would end up dying in huge numbers, leaving behind a few survivors for the future. Lost to neo-Malthusian pessimists was the fact that high death rates have traditionally raised human fertility. When death rates fell in the late twentieth century, family size also declined.[14]

Population According to Dyson and Hoyle

Fretting over population growth may seem far removed from astronomy. Such was not the case. In 1960, *Science* published "Search for Artificial Stellar Sources of Infrared Radiation" by Freeman Dyson (b. 1923), physicist and a fellow of the Institute for Advanced Study in Princeton. In this influential article, Dyson combined Malthusian fears with speculation on extraterrestrial engineering. Starting with the truism of humanity's high technology being rather recent, Dyson took for granted the superiority of older extraterrestrial civilizations and their expansion to "the limits set by Malthusian principles."

In handling runaway population, what could extraterrestrials do? Dyson later admitted his debt to Olaf Stapledon's *Star Maker*, in which future beings tap the entire energy of their star. Dyson noted that a Jupiter-sized planet could be disassembled and its matter used for a shell around a star. The shell could trap the star's energy and be made "comfortably habitable," producing both living space and energy. According to Dyson, this super-engineering project gave astronomers a second avenue of exploration. Apart from radio signals, astronomers could search for infrared radiation from the shell, which Dyson described as "a dark object."[15]

Was Dyson serious? Advanced extraterrestrials must have devised simpler solutions to population growth. It may be, though, that Dyson's spheres were to be taken with a grain of salt. He was using his status as a leading intellectual to warn humans of their overactive libidos. Maybe an explanation lies in Dyson's biography. Described as the best mathematician in England at the age of twenty-three, he immigrated to the United States where he made a reputation as a brilliant maverick, unafraid of controversy. In recent years, although a liberal in politics, he has been a climate skeptic, congratulating the Chinese for burning coal and industrializing. Dyson has liked to throw out provocative ideas, and his "Dyson spheres" may have been simply another hypothetical for discussion. Whatever his motives, Dyson did present an alternative SETI strategy: searching for infrared radiation instead of radio signals.[16]

Cambridge University astronomer Fred Hoyle (1915–2001) shared Dyson's unease with human breeding. Famous for coining the term "Big Bang" to show his rejection of the theory that the universe had a discrete beginning, Hoyle had lectured at Cornell University in the early 1960s, and population was very much on his mind. Like Malthus, Hoyle feared that the human race could not control what he called its "excessive reproductive vigor," but unlike Malthus, who predicted mass rioting, Hoyle viewed the future from an engineering perspective. According to Hoyle, society had to increase in complexity to serve the expanding hordes of humanity. The prognosis was grim, for like an enormous machine, which fails because it depends on millions of synchronizing parts, an overly complex society at some point collapses. After the debacle, Hoyle believed, the smartest and the more cooperative, presumably intellectuals like himself, would create a new society.[17] As with Dyson, the question arises whether Hoyle was serious or merely preaching to the masses. One thing was certain, though: as

Morrison's career demonstrated, a number of astronomers were queasy about the future and willing to stray from their field of expertise.

By the 1950s, the average American was content, if not smug, but many intellectuals were depressed. Unlike their forebears of the nineteenth century, they did not believe that common sense, intelligence, and science guaranteed human progress. They saw a dismal twentieth century, marked by overbreeding, world wars, genocide, and Hiroshima. Although the world wars and the Holocaust were behind us, demographic disaster and atomic bombs seemingly made humanity's suicide less the stuff of science fiction than a future reality.

It would have been comforting to know that mankind could survive. Perhaps, beyond Earth, intelligent beings had faced the same challenges and had overcome them. Finding out would not hurt. SETI would have two parents. Born of the same curiosity that had driven explorers throughout history, it was also the child of pessimism, a despair that made some very intelligent scientists hope that solutions lay in distant worlds and strange beings.

4

The 1960s

The Quest Begins

Strictly speaking, the SETI age began with Philip Morrison and Giuseppe Cocconi's 1959 article in *Nature* and Frank Drake's Project Ozma the following year. A case can be made, though, that the Age of SETI began when Karl Jansky accidentally discovered interstellar radio waves.

In 1931, Karl Jansky (1905–50), a physicist with Bell Laboratories, had the assignment of understanding why static interfered with radio transmissions. He identified thunderstorms as one source of static, but a faint steady hiss proved mysterious. After much study, Jansky realized that a natural radio signal from the galaxy's center was hitting Earth. Although Jansky wanted to continue his research, Bell Labs saw no immediate use. Jansky was reassigned, and he never again worked on interstellar radio waves.

After World War II, the implications of Jansky's discovery began to be appreciated. Jansky had laid the groundwork for radio astronomy, a new form of interstellar exploration that supplemented traditional optical telescopes. Because objects in space emit radio waves that slice through interstellar gas and dust, astronomers with radio telescopes could "see" these objects, even though Earth's atmosphere spoils an optical view.[1] Jansky's work also implied that intelligent beings could send artificial radio waves to Earth.

Philip Morrison Launches the SETI Age

At Cornell, in the 1950s, while Morrison was specializing in the origin of cosmic rays, he wondered whether extraterrestrials were using gamma rays, a particular form of cosmic rays, for interstellar communication. Along with his colleague Giuseppe Cocconi, who had studied under Enrico Fermi (1901–54) in Italy, Morrison looked into this possibility and, after some thought, rejected it. The two physicists decided that extraterrestrials were more likely using radio waves to communicate.

The September 1959 article in *Nature* resulted. In "Searching for Interstellar Communications" they argued that radio waves, traveling at the speed of light (approximately 186,000 miles per second), can bridge the vastness of space and that extraterrestrials might be using radio signals to contact Earth or to communicate with each other. Whichever the case, Morrison and Cocconi pointed out, Earth's radio telescopes might pick up an extraterrestrial radio message.

Natural radio waves aside, Morrison and Cocconi were hoping that extraterrestrials were transmitting information through artificial radio waves, as Earthlings do. The focus on radio waves implied a rejection of other electromagnetic waves, not only gamma rays but also microwaves, infrared, visible light, ultraviolet, and X-rays. Although all electromagnetic waves travel at the speed of light, radio waves have an advantage: Earth's atmosphere is less likely to block or distort them. Morrison and Cocconi reasoned that extraterrestrials know this.[2]

A focus on radio waves alone was insufficient; Morrison and Cocconi had to identify an extraterrestrial's radio frequency. A frequency refers to an electromagnetic wave's crests and troughs and the number of times that a crest passes a point every second. The radio part of the electromagnetic spectrum alone contains billions of different frequencies.

Which frequency were extraterrestrials using to contact Earth? Edward Purcell (1917–97), Nobel laureate in physics in 1952, used the analogy of two friends trying to meet without a prearranged meeting place. The two friends might independently decide to meet at a famous landmark. New Yorkers, for example, might choose Grand Central Station.

Morrison and Cocconi reasoned that the aliens had chosen a frequency whose background noise from both Earth's atmosphere and the galaxy was very low. This radio landmark was the 1,420 MHz frequency, which has a wavelength of twenty-one centimeters. This was the radiation given off by hydrogen atoms on the electromagnetic spectrum. Hydrogen has the added advantage of being the most common element in the universe.[3]

Otto Struve Defends His Turf

The immediate effect of Morrison and Cocconi's article was to provoke a competition. Modern scientists are more similar to the explorers of old than they care to admit. Columbus and the Spanish conquistadores allegedly labored for gold, God, and glory. Most modern scientists do not care about God, although many of them (like Philip Morrison) are passionate for its modern equivalent, which is saving the world instead of souls. Gold takes the modern form of grants, promotions, and the lucrative patents that reward scientific discovery. All scientists want the glory: the recognition from peers, entry into the history books, the big prizes, and the top sign of recognition, the Nobel Prize. These baubles of success invariably go to the trailblazer.

Morrison and Cocconi caught the attention of Otto Struve (1897–1963), director of the National Radio Astronomy Observatory in Green Bank, West Virginia. Like many American scientists of his day, Struve was a refugee—but not from Nazi Germany. Struve was a Russian of German descent who had fought the Communists after the 1917 Revolution. Escaping to the United States, he took up the family trade, having come from a long line of astronomers, and prospered. Struve's observatory had recently installed a radio telescope with a diameter of eighty-five feet, and for nearly a year it had been preparing Frank Drake's Project Ozma. Suddenly, Morrison and Cocconi published their article, and the implications were enormous. Cornell University might upstage Drake and the Green Bank Observatory. After all, a permanent place in history awaited the first astronomer to contact extraterrestrials.[4]

Struve was highly respected by his fellow astronomers, and this respect allowed Struve some maverick views. Whereas many of his peers doubted the existence of extra solar planets, Struve believed that billions dotted the galaxy. In addition, Struve held that intelligent beings lived on many planets and advised astronomers to consider biological activity as well as the laws of physics when analyzing the workings of the universe.[5] Struve did not suffer from a Percival Lowell hangover.

Scheduled to lecture at MIT in November 1959, Struve changed the subject to Green Bank's Project Ozma. After noting that Earth's radio telescopes could detect radio signals from advanced extraterrestrials, he revealed that in early 1960 the Green Bank telescope would focus on two nearby stars, Tau Ceti and Epsilon Eridani, both less than twelve light-years from Earth.[6]

Frank Drake and Project Ozma

Struve's announcement of Project Ozma also marked the coming-out for the twenty-nine-year-old Frank Drake, a Harvard PhD in astronomy, who had recently joined the Green Bank staff. A serious astronomer whose prematurely white hair gave him an air of gravitas, Drake had a poorly concealed romantic side. He called his search "Project Ozma" after Ozma, the princess in Frank L. Baum's *Ozma of Oz*. Like Baum, Drake was also "dreaming of a land far away, peopled by strange and exotic beings."[7] Although he failed to detect any extraterrestrial radio signals, Drake has the honor of being the first to make a SETI search.

The Green Bank radio telescope had been deliberately placed in a remote West Virginia valley to have the surrounding Allegheny Mountains block radio and television broadcasts. Although a great place for astronomers, according to Drake, Green Bank was isolated and a boring place to live. One snowy day while lunching at "a greasy spoon diner" nicknamed "Pierre's," Drake casually suggested to colleagues that the Green Bank telescope should listen for alien signals. His colleagues agreed and Drake later recalled that "as the last greasy french fry was washed down by the last drop of Coke, Project Ozma was born."[8]

Independently of Morrison and Cocconi, Drake had also grasped the potential of the hydrogen line frequency. He calculated that the Green Bank radio telescope would detect signals from extraterrestrials no more than ten light-years distant, if their radio transmitters were as powerful as Earth's. He later wrote that he had "resolved to keep Project Ozma a secret from start to finish, so as to avoid publicity and interference from the press." Drake knew his work was unconventional; he feared "that other astronomers might scoff at it." If criticized, he could claim he was not wasting money; Project Ozma's equipment also served conventional radio astronomy. The Percival Lowell curse loomed large. Once Drake realized, though, that Morrison and Cocconi had let SETI out of the closet, it was time to speak out.[9]

Drake and Morrison differed in explaining their interest in extraterrestrial life. Although Morrison had read science fiction, especially H. G. Wells, he insisted that it was irrelevant. According to Morrison, science fiction had merely made the culture (and presumably himself) aware of extraterrestrial life.[10] Drake, however, saw a religious origin to the SETI quest. He has claimed that an "extensive exposure to fundamentalist religion" can be a common thread connecting "people who have been active in SETI." His parents were strict Baptists who sent him to a Sunday school in which archaeologists from the Oriental Institute of the University of Chicago used field work in the Middle East to support the Bible. Drake visited the science museum in Chicago, and at an early age he learned enough of science and astronomy to see a conflict between Sunday school and "what the universe seemed to be." He eventually rejected the Sunday school version.[11]

Drake has not explained the exact connection between religion and SETI. Interestingly enough, the same *New York Times* article that reported Struve's lecture at MIT also reported Drake's questions for extraterrestrials. He wanted to ask how "to prevent cancer and heart disease; how to prolong life; how to control the energy of the fusion process in the hydrogen bomb for industrial power; how to develop man's creative potential; and, above all, whether—and if so how—the planetary society had managed to build a culture at peace in which each individual lived a full physical and spiritual life."[12]

Besides being good public relations for Project Ozma, these questions reveal that Drake was unwittingly projecting a heaven of sorts onto an extraterrestrial civilization. Drake was resembling his Sunday school teachers, becoming an archaeologist of sorts who eschewed biblical realms for the new Holy Land of space. Like his Baptist teachers, Drake sought comfort and meaning. Unlike his teachers, though, he eschewed scripture, searching the stars instead for answers.

On April 3, 1960, the *New York Times* reported that Project Ozma was ready. Five days later, at 3 a.m., Drake and two student assistants started the search. By the standards of later SETI searches, the equipment was simple, almost primitive, a simple receiver with one channel. A bit nervous, and wondering whether they were foolish or on the cusp of "a historic moment," they first aimed the telescope at Tau Ceti. At noon they switched to the other targeted star system, Epsilon Eridani. Within five minutes they heard a brief but strong pulsed signal, an artificial signal. Although excited, Drake was cautious; the signal could have been terrestrial. Drake later learned that the mysterious signal came from a military plane, almost certainly a U-2 spy plane, that the United States was secretly flying over Communist countries. When reporters found out about the signal, they quizzed Drake, who knew little and refused to speculate. He paid a price for his professional silence. Flying-saucer buffs assumed he was lying and invented an alternative scenario in which Drake had contacted extraterrestrials and was conspiring with the U.S. government to hide the news from the public.[13] Drake was also silent about the receiver malfunctions and the paucity of data he received—about two hours' worth. If nothing else, there was an excuse for not contacting ET.[14]

Project Ozma had cost little, about $2,000. Lack of money did not end it; Drake had merely used up his allotted time on the Green Bank telescope. Other astronomers were waiting to use the equipment. If silence is approval, the public approved of Project Ozma. Drake received no angry phone calls or letters. What else could be expected from a public familiar with ufology and science fiction?

Project Ozma had been an optimistic gamble, a throw of the dice in the cosmic crap game. Drake knew well that the universe is huge beyond imagination and contains trillions of stars, yet he was betting that extraterrestrials lived nearby and shared our curiosity. For good measure, Drake was betting these extraterrestrials were using radio, just like twentieth-century Earthlings, and had chosen from billions of frequencies the same one that he had chosen. Even more optimistic was expecting a cancer cure and advice for achieving world peace. To believe in cures and advice was the seduction of science fiction, whose aliens often have humanoid bodies and think like humans.

Yet the search for extraterrestrials had to start at some point in human history. One might as well start with the easy route and move on when it fails. Optimism is necessary for progress, since obsessing on pitfalls and difficulties means never acting. After all, Columbus had been a supreme optimist, ignoring the experts who rightly told him that he would never reach Asia by sailing west. Although Columbus did not find Asia, he did find something totally unexpected. In 1960, who knew what Project Ozma could have discovered?

In launching the SETI era, Drake, along with Morrison and Cocconi, invented a novel space exploration that eschewed spaceships and astronauts.[15] They also in large part laid down SETI's agenda. Their suggestion that extraterrestrials were using electromagnetic signals, especially radio, has dominated SETI thinking ever since. Also important was their belief in a "magic" frequency, the one ET was likely using. For the next two decades, until spectrum analyzers allowed searching millions of frequencies, SETI researchers either accepted the 1,420 MHz frequency or made a case for a better "magic" frequency.

In the staid world of astronomy, Project Ozma was something different. Like all novelties, it ran the danger of being derided or ignored. Much of the astronomical community needed convincing, or else Project Ozma would end up sterile, destined to leave no progeny. A first step in gaining respect for SETI (the term was not yet invented) was showing that Project Ozma had been sensible instead of foolish, a cutting-edge experiment rather than a lost weekend for reputable scientists who had read too much science fiction.

In seeking to gain respect, Drake, Struve, and Morrison could present several arguments. They could point out, as they believed, that contact would greatly benefit humanity, or they could assure pessimists, frightened by the *War of the Worlds* paradigm, that aliens were peaceful. Still another tactic would be defining contact as a step in the

continuum of human evolution. These arguments are still made to justify SETI.

The best short-run argument for continuing SETI was showing that
the next Project Ozma might succeed. It had to be shown that space
was full of extraterrestrial civilizations, that space was less like an empty
desert and more like a crowded marketplace in which ideas, if not necessarily goods, were exchanged. If intelligent creatures populated the
universe, SETI would not return astronomy to the embarrassing days
of Percival Lowell. It would instead evoke memories of the Portuguese
explorers of the fifteenth century who believed in an all-water route to
India, patiently searched, and found it.

Although some astronomers had privately discussed the possibility
of extraterrestrial civilizations, the discussion needed to come out of
the closet. The search for extraterrestrials needed the approval, or at
least the tolerance, of the scientific establishment, or, to be specific, its
governing and accrediting agencies. One such agency was the National
Academy of Sciences' Space Science Board, formed in 1958 to lay out
American goals in space.

The Green Bank Meeting

In December 1961 the Space Science Board convened a meeting at
Green Bank to discuss extraterrestrial societies. The meeting was not
publicized. With millions of Americans believing in flying saucers, a
publicized meeting would draw reporters from both the reputable and
the not-so-reputable press, all of them a nuisance. Complementing the
press would be self-styled "experts," armed with tales of alien experiences and demanding a hearing.[16]

J. P. T. Pearman of the Space Science Board organized the meeting.
Working with Drake, he selected eleven participants, achievers in their
chosen fields and believers in extraterrestrial life. Drake, Morrison,
and Cocconi were of course invited. Cocconi declined, having already
returned to Europe to work at CERN, the European Organization for
Nuclear Research. He had lost interest in SETI and refused to discuss
it, claiming that he had not kept up with the literature.

The other eight invitees (including Pearman) did attend the Green

Bank meeting. Two were businessmen. Dana W. Atchley Jr., president of Microwave Associates, Inc., had supplied Project Ozma with a parametric amplifier, which had greatly increased the receiving sensitivity of the Green Bank radio telescope. Bernard "Barney" Oliver (1916–95), vice-president of development at Hewlett-Packard, was visiting Washington, D.C., when he heard that Drake was conducting Project Ozma. An avid science-fiction reader in his youth, Oliver was so excited that he flew to Green Bank in a private plane to see for himself. He was destined to be a giant of SETI's first generation.[17]

The other invitees were scientists and researchers. Melvin Calvin (1911–97), professor of chemistry at UC Berkeley, had shown how life might arise from inanimate matter. He was convinced that Darwinian evolution was common throughout the universe. John Lilly (1915–2001), a medical doctor, might have appeared out of place among astronomers and physicists, but he had much to add. His recent publication *Man and Dolphin* claimed that dolphins were very intelligent and, by implication, that human intelligence was no freak event.

Three more astronomers rounded out the ten. Su-Shu Huang (1915–77) of Princeton University's Institute of Advanced Study had speculated on planets that might harbor life. Carl Sagan was not yet thirty and in heavy company, but throughout his life he had never lacked self-confidence. Armed with an extensive knowledge of biology, Sagan perhaps believed in extraterrestrials more than the other Green Bank men. Finally, there was the host: Otto Struve himself.

The Green Bank Ten held that humanity would gain much from extraterrestrials. The Brookings Institution think tank disagreed. Its report submitted to NASA in December 1960 had painted a pessimistic picture of contact. According to Brookings, self-satisfied human civilizations disintegrated when they confronted superior societies. The report warned of a similar outcome following contact with extraterrestrials. It could be inferred from the Brookings report that humanity was not ready for extraterrestrials and that searching for them was foolish, if not dangerous.[18] The Green Bank Ten dealt with the Brookings report very simply: they ignored it.

The Green Bank Ten also ignored Harlow Shapley (1885–1972), professor emeritus of astronomy at Harvard University. In June 1960, commenting on Project Ozma, Shapley had conceded the possible presence of nearby intelligent creatures but dismissed as "amusing" the thought that they were trying to contact Earth. Shapley pointed out that the vast distances of space discouraged an effective dialogue. Radio signals from Tau Ceti and Epsilon Eridani, the star systems that Drake hoped to hear from, would take ten years to arrive, and Drake's reply would require another ten years. Shapley doubted that extraterrestrials waited this long for a conversation.[19] Shapley could have added that conversations with more distant extraterrestrials would take hundreds, if not thousands, of years. One might discern an implied slap at Drake, who had given the impression to the *New York Times* of an instant conversation with aliens.

Although Shapley was prestigious in astronomy circles, SETI survived his sour judgment. Contacting extraterrestrials was an idea that resonated beyond the small world of astronomy. Apart from tapping into religious sensibilities, it appealed to science-fiction buffs and to the millions more who had watched Hollywood space operas. Besides, Drake had a knack for publicity. The term "Project Ozma" had a touch of genius. It made Drake's scanning of Epsilon Eridani and Tau Ceti stand out from other scientific investigations. By giving the media and fellow astronomers a concrete image, taken from a popular children's book, Drake made his SETI search something special, even if it failed, which, after all, it did.

Drake's greatest public-relations coup was inventing an equation that bears his name. At the Green Bank meeting, when presenting the agenda, which was to estimate the number of extraterrestrial civilizations, Drake did not randomly list topics on a blackboard. Instead, he linked them in equation form, inventing the so-called Drake Equation, which, often repeated and debated, has become a SETI staple.

$$N = R\, f_p\, n_e\, f_l\, f_i\, f_c\, L$$

N: the number of detectable extraterrestrial civilizations
R: the rate at which new stars form in the Milky Way galaxy each year
f_p: the fraction of our galaxy's stars that have planets

n_e: the number of planets in each star system with conditions conducive to life

f_l: the fraction of planets in n_e where life emerges

f_i: the fraction of planets in f_l that have intelligent beings

f_c: the fraction of planets with intelligent beings that are capable of interstellar communication

L: the lifetime of a civilization that communicates to other worlds

Its beguiling simplicity has given the Drake Equation tremendous staying power. It has captured the public fancy, even appearing on T-shirts, truly a sign of iconic status. Yet the Drake Equation can be easily criticized. It is not a real equation, because it multiplies unknowns whose values can vary greatly, thereby permitting any desired result. One could argue that the equation's variability makes it closer to the social sciences than to astronomy, especially since its last parameters are sociological. The equation calls to mind the social science conceit of using mathematical symbols to feign the rigor of the natural sciences. The equation is nonetheless useful. A convenient starting point for a SETI discussion, it identifies the dark areas needing more research if extraterrestrials are to be found.

During their three-day meeting, the Green Bank Ten discussed the equation's terms and assigned values, which had to be generous in order to inflate N. A huge N (number of extraterrestrial civilizations) increased the chances of intelligent life on a nearby planet contacting Earth. Unfortunately, precision was not possible. In 1961, and still today, the galaxy is like a huge jigsaw puzzle with many missing pieces. The known pieces tantalize and allow astronomers to speculate hopefully, even wildly, about the unknown parts of the puzzle. The Green Bank Ten knew their astronomical ignorance allowed great leeway in assigning values to the equation. They discussed, debated, and finally, on the basis of their limited knowledge of the terms' true values, assigned the most favorable numbers that would not seem ridiculous.

The first variable, R, the rate of yearly star formation, was given the value of 1. The astronomers at the conference agreed that every year a new star is formed in our galaxy.

Less certain was the value of f_p, the fraction of the stars in the Milky Way galaxy surrounded by planets. The Green Bank Ten knew of only one star with a planetary system, and that is our sun. Drake suspected that the gravitational tugs of unseen planets explained the irregular movements of certain stars. Although right, he had to wait until the 1990s for confirmation. Today there are more than 4,600 candidate exoplanets (planets beyond the solar system), but in 1961 only educated hunches were possible. The Green Bank Ten decided that as many as half of the stars or at very least a fifth might have planets.

As for n_e, the number of life-friendly planets in each star system, everyone knew with certainty of at least one planet (ours) with life. Su-Shu Huang noted the chance of some kind of life on Mars and Venus. Today, the prospects of Martian and Venusian life have diminished. The Russian Venera probes of the 1960s and 1970s saw a Venus that fit the traditional view of hell: a surface temperature of over 700 degrees Fahrenheit. The Viking II lander of 1976 found no life on Mars (optimists are still hoping, though, for subsurface Martian life).

In 1961, the Viking and Venera probes had not yet taken place. Assigning n_e a value of 1 could be regarded as conservative, especially after Carl Sagan spoke. Obsessed with extraterrestrials from youth, Sagan readily adopted any hypothesis pointing to their existence. He argued that a planet far from its sun could be warm if covered with greenhouse gases. Sagan's intervention resulted in the value of n_e being raised to as high as 5.

Since the potential for life does not necessarily indicate its presence, the f_l variable examined the question whether planets with the right stuff for life produce it. The answer to this question leads to still another question: the big one on how life began. According to astrobiologist and cosmologist Paul Davies, three "philosophical positions" on life's origins are possible: "(i) it [life] was a miracle; (ii) it was a stupendous improbable accident; and (iii) it was an inevitable consequence of the outworking of the laws of physics and chemistry, given the right conditions."[20] The first position does not belong in science. The second position is popular with biologists, who see life's emergence as unlikely. The third position appeals to life-origin optimists, especially the SETI community, whose universe teems with life.

Carl Sagan and Melvin Calvin were optimists, maintaining that the proper conditions automatically produce life. They were referring to the work of Harold Urey (1893–1981), his graduate student Stanley Miller (1930–2007), and Calvin himself. In 1950, Calvin had produced organic compounds from inorganic substances, but he did not produce amino acids. This achievement belonged to Urey and Miller. At the University of Chicago in 1953, they conducted the classic experiment on life's origins. Using electrical sparks to simulate lightning, they derived organic chemicals as well as amino acids from methane and ammonia gases, inorganic chemicals found throughout the universe. Urey and Miller were far from creating life—besides, it is no longer believed that the early Earth was abundant with methane and ammonia—but they did produce life's building blocks and implied that life appeared naturally.[21]

Calvin was very much a true believer, convinced that the chemical origin of life and Darwinian evolution were common throughout the universe. Sagan fully agreed. Both men were preaching to the choir. The Green Bank Ten decided that the conditions for life invariably beget it. The variable f_l received the value of unity, 1.

Although possibly common, extraterrestrial life might be dumb, no more advanced than terrestrial bacteria or slugs. The Green Bank Ten would have welcomed a universe of slugs, provided the ugly things were little Einsteins. On the other hand, extraterrestrial slugs as dumb as ours would have been depressing, since a stupid universe would make SETI somewhat pointless. Not surprisingly, the f_i variable, the frequency of intelligence, triggered a lively discussion. As always, the Green Bank Ten had only a single model to consider, Earth, which in several billion years has generated billions of species but only one with intelligence—or maybe not so.

John Lilly had a message that his Green Bank colleagues enjoyed hearing: humans were not the only intelligent creatures on today's Earth. Lilly studied dolphins for the U.S. government at the Communication Research Institute in the Virgin Islands, and he had concluded that dolphins were very intelligent, cooperative, and spoke a complex language, although he admitted to not understanding it. Lilly enthusiastically described his work, and the Green Bank Ten wanted to believe.[22]

If Lilly was correct, Darwinian evolution had produced a second intelligent species on Earth, thereby increasing the odds of intelligence arising elsewhere in the universe. There was still another dimension. If Lilly's findings were valid, he belonged in the same continuum as Copernicus and Darwin. When Copernicus had proposed the heliocentric theory, he displaced humans from the center of things, destroying certitudes predating the classical Greeks. Darwin further demoted humans when he linked them to the so-called lower animals. Now Lilly was taking another step in deflating human egos by demonstrating that *Homo sapiens* did not have a monopoly on terrestrial intelligence. Lilly might well have been the superstar at Green Bank, greater than Melvin Calvin, who received word during the meeting that he had been awarded the Nobel Prize in chemistry. Morrison, Drake, and the rest were potentially hobnobbing with one of the greats.[23]

To speak of dolphin intelligence begs the question of what constitutes intelligence. Humans often define intelligence in terms of their own behavior, especially the transmitting over generations the insights learned from experience.[24] Although animals also survive through adaptation, human additive ability accelerates over time, whereas animals change so slowly that they are practically changeless. No evidence exists that today's dolphins act differently from the dolphins of five thousand years ago, when recorded history began. In the short run, dolphins do not seem to learn. Notorious for chasing tuna boats, dolphins die when they entangle themselves in tuna nets. If dolphins were so smart, the species would have figured this out.

Isaac Asimov (1920–92) was a Boston University professor of biochemistry, better known for his science fiction. He defined intelligence as the ability to create a complex technology.[25] Another definition of intelligence has creatures creating symbols useless for survival—art and music, for example. Maybe dolphins have a literature, oral epics they sing to each other in their unique "language" of whistles. Hence the ongoing research into dolphins and the hope of detecting that literature: a dolphin *Iliad*, an epic of heroic cetaceans and their battles with oversized sharks. Dolphin intelligence aside, dolphin research is attempted cross-species communication, a preparation for contact with ET.[26]

Another sign of intelligence has been tool making, a yardstick that paleontologists use to assess prehistoric hominids' coping skills. Unfortunately, the tool-making requirement excludes dolphins, as Morrison noted at the conference. While Lilly's rhapsody on dolphin intelligence was beguiling his colleagues, Morrison pointed out that dolphins live in water, make no tools, and possess no radio technology.

Besides, the story of evolution shows that intelligence is not necessary for survival. Calculus is quite foreign to rats and roaches. All the same, these vermin emerged before humans and will probably outlast them. Furthermore, intelligence has led to the SETI community's great fear: the suicide of humanity.[27] These thoughts, though, are heretical in SETI circles. The Green Bank Ten wanted to believe in an intelligent universe. In assigning a value to f_i, the fraction of life-bearing planets with intelligent life, they gave an optimistic value of one, confidently stating that life eventually gets smart.

Intelligent extraterrestrials were insufficient for the Green Bank Ten. They well knew that Newton and Galileo were ignorant of radio signals. Even for part of the twentieth century, humanity was ignorant of the radio universe. If extraterrestrials had a technology similar to Earth's during World War I, their world would boast aircraft and submarines, even an Einstein writing on relativity, but these extraterrestrials would have no television or radar to leak electromagnetic signals into space. Nor could they send dedicated radio signals to galactic neighbors. Without radio astronomy, ET is silent to the universe. The f_c term, the number of extraterrestrial civilizations communicating across the stars, was tough to assign any value to.

Morrison took comfort in the fact that China, the Middle East, and Mesoamerica, though isolated from each other, all developed civilizations. Morrison was assuming that any civilization, given enough time, eventually produced a high technology. In Earth's history, though, only the Middle East, which gave rise to the West, produced an civilization that was capable of interstellar communication.

Even in the Western world, technological progress was no certainty, for it required the confluence of numerous cultural streams. If just one was absent, would the entire process be short-circuited? Or would intelligence and curiosity make that absence a temporary detour in the march to high technology? The problematic role of Christianity in the rise of Western science is illustrative. According to Sagan, although Christianity killed off classical Greek science, it only temporarily slowed the march of progress.[28] Later in European history, the travails of Galileo suggest that Christianity remained a roadblock, if not a mortal threat, and without its weakening in the seventeenth and eighteenth centuries modern science would not have developed.

Another view of Christianity is possible. Christianity was not an obstacle but a necessary condition in the rise of Western science. Christians believed in a rational god whose creation was knowable and safe because their rituals trumped demons and other mysterious forces in nature. All this encouraged examining the natural world. This scenario acknowledges the persecution of Galileo and dismisses it as irrelevant. In other words, without Christianity to create the setting and the preconditions, there might not have been a Galileo, a Newton, or even a Darwin.[29]

Whether Christianity was necessary for the rise of science is beside the point; the issue here is not Christianity but the contingency of high technology, a hypothesis the Green Bank Ten ignored. What could be expected of scientists who pontificated on the growth of civilizations without consulting a historian? Even so, a historian's absence probably did not matter for the Green Bank Ten.

In debating the value of f_c, Calvin was less guarded than Morrison. Calvin argued that intelligent extraterrestrials could see, hear, and touch, even if their versions of eyes, ears, and hands substantially differed from ours. He was convinced that intelligent, technologically sophisticated creatures would eventually discover electromagnetic radiation. Nonetheless, the Green Bank Ten admitted they were speculating. Even if Calvin was correct, high intelligence does not imply a multicultural curiosity about interstellar neighbors. ET might be both advanced and self-centered, arranging his rocks, little concerned with the rest of the universe.

If extraterrestrials are not trying to contact us, could they at least be leaking their version of television signals or military radars into space? Will Earth's radio telescopes detect the alien equivalent of the soccer World Cup? Although the Green Bank Ten were optimistic, in no way could they assign a value of "1" to f_c. Before the twentieth century, humans could not signal the stars. Extraterrestrials could be far behind, at best with technology matching the early twentieth century on Earth and at worst locked in a Stone Age, chiseling fine arrowheads but millennia from radio. The Green Bank Ten were hoping, however, to find older star systems whose inhabitants evolved long before humans and whose technology was far superior. They gave to f_c a compromise value ranging from 0.1 to 0.2.

Finally, the Green Bank Ten considered L, the lifetime of a communicating civilization, the key variable of the entire discussion because of the fear for human survival. Hiroshima and Nagasaki had traumatized Morrison as well as many other scientists. Their spirits sank further in the 1950s when both the Americans and the Soviets stockpiled atomic weapons. The new decade of the 1960s brought little hope. The Soviet leader Nikita Khrushchev had a weakness for saber rattling, and President Kennedy was willing to accept the challenge.

Also troubling the Green Bank Ten was the thought that avoiding nuclear war was not enough; humanity's overactive libido also had to be checked. The Green Bank Ten saw their fellow humans procreating at a rate far beyond the capacity of Earth's resources and feared that this might bring about a new Dark Age, if not extinction itself.

With nuclear and Malthusian disasters troubling the Green Bank Ten, L was their key variable, especially if the other variables were seen optimistically as resulting in a product of 1, shortening the Drake Equation to N = L. Since the Green Bank Ten wanted a large N in the Drake Equation, a large L was also necessary. This line of reasoning was not idle speculation fueled by curiosity about extraterrestrials. The Green Bank Ten assumed that the life spans of extraterrestrial civilizations foretold humanity's future. If extraterrestrials survived their versions of lethal technologies and population explosions, their civilizations would endure for millions of years, giving humanity reason to hope. On the other hand, according to the Green Bank Ten, if an extraterrestrial civilization did not meet its challenges, its L was one thousand years or less. The Green Bank Ten concluded that the number of advanced civilizations ranged from as few as forty to as many as fifty million.[30]

The Green Bank Ten were obsessed with the present. They regarded the late twentieth century as the key period in history because humanity was entering the high technology stage. If humanity got it right in the upcoming decades, the human species was destined to near immortality. If not, humanity would be a failed species. With so much being asked of their generation, the Green Bank Ten were justifying political activism. Philip Morrison could not have asked for more.

To believe that we can learn from extraterrestrials assumes that their psychology is similar enough to ours that we can relate to their "wisdom." If some extraterrestrial civilizations meet this requirement, a second problem presents itself. The Green Bank Ten was assuming that a civilization can achieve perpetual peace, a political end of sorts, yet at the same time continue to develop its technology and science—a bifurcated culture of stasis and change. Underlying this assumption was a belief in ethical progress, the idea that older extraterrestrial races evolve morally. This may be so, for the older races that did not evolve morally and renounce war probably did self-destruct, leaving behind only the peaceful. On the other hand, all this may be wishful thinking. The Green Bank Ten wanted to believe in a universe of long-lived civilizations, a paradise in outer space of science and peace, because this utopia implied a glorious future for humanity.

The Green Bank Ten missed the real meaning of Lilly's assertions about dolphin cognition. Wanting an intelligent universe, they saw Lilly's research pointing in that direction. They ignored Morrison's insight that dolphins, even if intelligent, cannot receive or send radio signals: only creatures with intelligence similar to ours are relevant to SETI. Besides, if given a cold hard look, Lilly did little to encourage SETI. Of the billions of species that Earth has produced, dolphins and humans are the only ones allegedly intelligent, yet they cannot communicate with each other. The prognosis for human-extraterrestrial dialogue is not good.

The Green Bank Ten had not met, though, to conclude that alien civilizations were few or nonexistent and that searching was a waste of time. Not surprisingly, Lilly's tales of dolphin intelligence left them seduced. They dubbed themselves the "Order of the Dolphin" and awarded themselves a dolphin pin. Like vacationers who meet at a resort, they promised to meet again, but they never did. Their enthusiasm for Lilly turned out to be embarrassing. Other researchers could not duplicate his alleged communication with dolphins. Even worse, Lilly later claimed that he had spoken to extraterrestrials while taking psychedelic drugs.[31]

The Aftermath of the Green Bank Meeting

The Green Bank meeting might have impressed the lay world, whose information came from the media. In a feature story titled "Contact with Worlds in Space Explored by Leading Scientists," the *New York Times* on February 4, 1962, reported the conference's estimates of a universe full of extraterrestrial civilizations.[32] The public, conditioned by science fiction and ufology, would not realize that the estimates in the *Times* were flights of optimism.

Before the conference, Struve had written that Project Ozma had deeply divided the astronomy community. Some astronomers were enthusiastic; others dismissed Project Ozma as a publicity stunt.[33] Clearly, the Green Bank meeting sought to correct these negative impressions. By proposing a universe teeming with life, the conference was giving retroactive legitimacy to Project Ozma and justifying future searches.

The Green Bank Ten believed that they were laying out a new course for radio astronomy. This they were. Nevertheless, their political and social agenda cannot be ignored. Although they would have denied it, they stood in the same continuum as George Adamski and other ufologist oracles. Like Adamski, the Green Bank Ten were using extraterrestrials to protest humanity's possible suicide. It was irrelevant that extraterrestrials were unknown and perhaps nonexistent: they were crucial. Without them, Adamski was a blue-collar ice cream salesman. Without them, the Green Bank Ten were scientists straying beyond their field of expertise. Yet Adamski and the Green Bank Ten were crucially different. Adamski's labors resulted only in a minor fame. The Green Bank Ten could have hit the jackpot—conversing with ET.

In the history of SETI, the Green Bank meeting stands as the inaugural event in which the founding fathers of SETI laid out themes still around today. If it is conceived, however, as a launching pad for new extraterrestrial searches, the conference failed. In March 1962 a congressional subcommittee held a hearing on extraterrestrial life. Sir Bernard Lovell (1913–2012), director of the Jodrell Bank Observatory in Great Britain, testified that he believed that there was extraterrestrial life but questioned random searches such as Project Ozma. The search for extraterrestrials, Lovell told the legislators, required a long-term strategy, that is, a large number of dedicated radio telescopes. Lovell added that no nation would make this huge investment. Since Lovell was speaking at the height of the Cold War, he was probably referring to the Soviet Union. The implication was clear: if the Russians were not searching for ET, the money should be spent elsewhere. Dr. Harrison Brown (1917–86), geochemist at the California Institute of Technology, also spoke, and like Lovell he talked money. He doubted extraterrestrial legislators spent huge sums to send signals requiring countless lifetimes for a reply. Brown can be faulted for his anthropomorphism, that is, the assumption that extraterrestrials shared the human life span and allocated research funds through a legislative process. He did point out, like Lovell, the cost of a proper search. Neither Lovell nor Brown favored another Project Ozma.[34] As Drake later wrote, SETI was dead for the rest of the 1960s. Not until 1971 would an American astronomer attempt a follow-up to Project Ozma.

5

The 1960s

The Selling of SETI

The 1960s were optimistic times, the age of activism when it was widely believed that sex, drugs, and protest were the keys to a better world. Although less exuberant, scientists had their own version of euphoria. Almost giddy with the potential of that new tool of research, the computer, they saw artificial intelligence only a decade or two away. Joining machine-made intelligence was the almost certain prospect of creating life in the laboratory, stealing the glory from the god of Genesis.[1] Amid all this optimism, contact with extraterrestrials seemed almost a given. Astronomers of the 1960s, unlike their predecessors, had radio telescopes. Sooner or later, and SETI was betting on the "sooner," interstellar communication would stop being science fiction.

In the new millennium, science has yet to live up to these expectations. Despite advances, the laboratory has yet to generate life from inanimate matter, artificial intelligence has yet to fully replicate the human mind, and contact with extraterrestrials remains in the future. Many see themselves still alone in the universe, without a correspondent, singing to the empty sky like a coyote in the night, and wondering whether life is pointless.

In the go-go 1960s, Philip Morrison and Frank Drake were not alone in expecting ET to call. Extraterrestrials might have seemed a stretch for Ronald Bracewell (1921–2007), a professor of electrical engineering at Stanford who was studying magnetic resonance imaging and radio astronomy. Bracewell was fascinated by the possibility of extraterrestrials, and he believed that advanced extraterrestrials were likewise curious about us. In 1962 he wrote: "I believe we are on the eve of plugging in on the galaxy-wide communication network." He reasoned that ET would call because Earth might soon destroy itself, and alien anthropologists could not resist studying "a society near its peak." Bracewell managed to combine a wishful thinking about extraterrestrials with a profound pessimism that is sometimes seen in intellectual circles.[2]

ET did not call, and Bracewell along with Drake and Morrison kept on waiting. Besides, American astronomers were not searching for extraterrestrials in the 1960s; the few SETI searches that did occur took place in the Soviet Union. The decade after Drake's Project Ozma was perilous for American SETI, forced to justify itself. At a 1966 symposium in which Drake and Oliver spoke, Oliver was introduced (if not dismissed) as a science-fiction buff. He took great umbrage.[3] At the twelfth annual meeting of the American Astronomical Society in 1967, SETI was absent from the agenda.

From the start, Morrison and Cocconi realized that SETI needed legitimacy, given the scientific community's skepticism. In the final paragraph of their seminal 1959 paper they had denied that listening for extraterrestrial radio signals belonged in science fiction. Clearly, a persuasive case for SETI was necessary. Only this would induce public and private sources to finance proper searches. The SETI community's first task was getting attention.

The Space Race and SETI

Unfortunately for SETI, whenever Americans thought of space in the 1960s, first in their minds was neither knowledge nor curiosity, but politics—the need to outcompete the Russians. Although space was a Cold War battleground, this had not always been the case. Before 1957, believing it owned space, the United States had put space exploration on the slow track. In October 1957, American complacency abruptly ended when the Russians launched Sputnik, the first artificial satellite in space. The worldwide Communist propaganda machine went into high gear, proclaiming Soviet superiority in all things.

The Soviet Union laid down the challenge; the United States accepted it, and the space race was on. It began badly for the United States. On December 6, 1957, an American attempt to match Sputnik colossally failed when the Vanguard rocket exploded on the launch pad, shocking millions of Americans who watched with mouths agape on television. When the U.S. Army launched the tiny Explorer satellite in January 1958, the Russians countered with a larger satellite. A pattern was set that lasted till the early 1960s. American rockets exploded or, if they lifted into space, their payloads were minuscule by Soviet standards. Many Americans nervously wondered whether Soviet space spectaculars indicated that Communism was indeed the wave of the future. Even worse, in the Cold War climate of the day, the question was asked whether space was a launching pad to rain missiles onto a defenseless United States. A source of danger and a harbinger of national decline, space was no longer the plaything of the imagination.[4]

SETI Activists and the Worriers

Apart from NASA's obsession with placing satellites and astronauts into space, SETI had a second problem. An extraterrestrial signal might pose a greater danger than the Russian space program. In 1960 the Brookings Institution had warned that alien contact might unsettle the human race. Why tempt fate by seeking aliens? By contrast, both the American and the Soviet space efforts were safe; neither sought intelligent life. SETI apologists had to reassure the public that SETI was as safe as the space program, that contact with extraterrestrials posed no danger and, furthermore, would enhance humanity.[5]

No astronomer of the 1960s feared an invasion from the sun's planets. In fact, there was no evidence of extraterrestrial life of any kind in the solar system, although Drake, Morrison, Calvin, and others still hoped to find primitive life.[6] Along with other scientists, these SETI pioneers knew that H. G. Wells's genocidal Martians would forever remain fiction.

As for invaders from beyond the solar system, the SETI community has always insisted that potential invaders are too distant and cannot reach Earth. SETI advocates point to Einstein's theory of relativity, which holds that the faster an object moves, the greater its mass increases, increasing the energy required to move it. Getting a spaceship to near the speed of light, 186,000 miles per second, would require the United States' entire energy production over thousands of years. At the speed of light itself, the energy requirement is infinite, making light speed a barrier that cannot be surpassed. Yet at the speed of light, a trip from the nearest star, Alpha Centauri, still takes four years. Even today, more than fifty years after Project Ozma, a spacecraft traveling a tenth of the speed of light is beyond realization. If extraterrestrials from the other end of the galaxy should achieve the speed of light, they nonetheless require thousands of years to reach Earth; they cannot travel faster. Bracewell reassured the fearful: instead of navigating the vast distances of space, extraterrestrials will find raw materials and food closer to home.[7]

Ufologists are not impressed. Then and now, they dismiss the "experts," insisting that advanced extraterrestrials have broken the speed-of-light barrier and have arrived. They ingeniously note that if extraterrestrials had not solved twentieth-century engineering conundrums, they cannot be "advanced." Ufologists claim the proof of interstellar travel is everywhere: the UFOs and aliens that many people vow to have seen.[8]

Ufology aside, if a science-fiction-style invasion of Earth was improbable, radio contact was another matter. SETI advocates have always believed that advanced extraterrestrials mastered radio technology long before humans did. In a 1962 talk, Morrison speculated that radio-savvy aliens have been marking time, waiting for other races to enter the radio age.[9] Drake made the same point with greater force:[10]

> At this very minute, with almost absolute certainty, radio waves sent forth by other intelligent civilizations are falling on the earth. A telescope can be built that, pointed in the right place, and tuned to the right frequency, could discover these waves.

Like Bracewell, Morrison saw advanced aliens finding raw materials and living space closer to home. Information is something else. Morrison felt that our backward planet excited alien anthropologists and that radio was their best way to study us. ET is curious, not dangerous.[11]

Sebastian von Hoerner (1919–2003), an astronomer at the Green Bank Observatory, conceded that extraterrestrials might be evil but added that it did not matter. An interstellar conversation by radio takes a long time, its length depending on the extraterrestrial's distance from Earth. Our children, grandchildren, or distant descendants will receive ET's reply to our response, von Hoerner noted. Earthlings have the luxury of a lifetime, perhaps several or many, to digest ET's message— still another reason for not fearing extraterrestrials.[12]

Extraterrestrial Motives

Although extraterrestrials may be harmless, the question of *why* we should try to contact extraterrestrials arises. Pessimism over humanity's prospects was a poor selling point. Likewise, curiosity may have been a sufficient reason for the SETI community, but funding agencies, both private and public, needed practical justifications. History speaks to this. Profit obsessed Europe's golden age of exploration. Greed, not curiosity, made Spain and Portugal finance the voyages of Columbus and Vasco da Gama. The explorers themselves were obsessed with fame and riches. Before setting sail, Columbus drove a hard bargain. He demanded a title of nobility as well as a part of the profits if he succeeded.

It is not surprising that a preoccupation with costs and benefits has obsessed NASA. After all, the American space program went into high gear only when Soviet space feats had embarrassed the United States. America had to show the world that its technology and its very way of life surpassed the Soviet Union's. Apart from national prestige, which needed no explanation, NASA justified itself, and still does, by pointing to spin-offs: the applications of its scientific research to the civilian economy.

With SETI, the short term has always counted less than the end result. Although the SETI community has bragged over the years of its improved antennae, its new "listening" devices (multichannel receivers for radio telescopes), and its basic research in related fields such as the origins of life, its chief justification has been contact itself, the beginning of the real learning experience. The SETI community hopes that extraterrestrials will bestow a treasure trove of knowledge and wisdom, which will better humanity.

In effect, a payoff myth was invented. ET was defined as a Santa Claus, a bearer of intellectual, technological, and political gifts. A big payoff from a successful SETI search would justify asking NASA and private donors for funds and persuading astronomers to surrender radio telescope time for SETI searches. Drake ended his Project Ozma, it should be recalled, because his allotted time on the Green Bank Telescope had run out. Apart from justifying SETI to the scientific community, the government, and donors, the payoff myth has a second asset. It gives SETI an acceptable face to the public, which is conditioned by ufology and science fiction to believe in extraterrestrials and at the same time to fear them.

The goodies that ET would bestow were many. Before Project Ozma, Drake envisioned alien knowledge ending cancer and war. In 1963, A. G. W. Cameron (1925–2005), a Harvard astronomy professor, foresaw "valuable lessons in the techniques of stable world government."[13] Political science lessons, though, were mere bagatelles for Walter Sullivan, science reporter for the *New York Times*. In his SETI chronicle of 1964, *We Are Not Alone*, Sullivan saw humanity skipping "thousands or even millions of years" of research in order to eliminate "premature death, congenital deformity, insanity, as well as . . . prejudice, hatred and war."[14] Truly, no god of a traditional religion has been so generous. Indeed, ET is better, for unlike the God of the Adamist faiths, who condemns his creations to painful death, ET will cure cancer.

It can be argued that SETI provides some of the comforts of religion. Over the years, many Christians have found religious satisfaction in being busy and thus denying the devil his proverbial playground. SETI research likewise gives a transcendental validation to worldly pursuits. Although its astronomers do legitimate science, they are at the same time seeking answers to the big questions of life and thus satisfying a residual religious need. SETI workers have the comfort of believing that contact will better humanity, not only by providing practical advice for the present but also by learning if humanity has a destiny. SETI research gives a higher meaning to the mundane tasks of writing algorithms and checking computer data.

If extraterrestrials are truly a Santa Claus, their generosity needs an explanation. Walter Sullivan believed it was possible that a dying race sought an immortality of sorts by imparting its wisdom, much as humans seek continuity through their descendants.[15] In a realistic universe, though, a dying yet benevolent extraterrestrial race is not the norm, leaving open the question of why extraterrestrials would want to help humanity. If positions were reversed, humans would not help. Humans tend to selfishness; prolonged altruism is so unusual that lifelong altruists are considered the exception. Some Christian churches have always recognized this sad fact. They occasionally select exceptions to the rule of selfishness and declare them "saints," role models for the faithful.

In the SETI universe, though, advanced extraterrestrials differ totally from greedy, selfish humans; they are ethically superior. The proof is in their survival. If they had not overcome their versions of original sin and political pathologies such as Nazism, they would be extinct.[16] According to Bracewell, the older extraterrestrial races are happily in contact with each other. Borrowing Stapledon's fictional Interstellar League, Bracewell posited a Galactic Club whose members exchange information via radio, spreading their cultural wealth, their recipes for survival and joy. The motive was not only altruism, according to Bracewell, but also self-interest so that emerging civilizations like ours do not go off the track and endanger the galaxy.[17]

Fred Hoyle went a step further; he envisioned Earthlings tapping into a Galactic Library, a shared knowledge pool in radio space of perhaps a million extraterrestrial civilizations, and using this collective wisdom to avoid catastrophic war and overpopulation.[18] If the Galactic Club and its attached Galactic Library (a galactic Internet in today's language) hint of science fiction instead of hardheaded science, it is no accident. The term "Galactic Club" comes from Isaac Asimov's *Foundation* novels, in which future humans have spread throughout the cosmos, although not necessarily in harmony. Hoyle, besides being a Cambridge don and astronomer, wrote science fiction on the side.[19]

Whatever ET's recipe for survival was, Bracewell, Drake, and Morrison did not publicly speculate. The ever energetic Sagan did not share their reticence. His 1966 book, *Intelligent Life in the Universe,* contained several sentences on extraterrestrials that might have been dismissed as science fiction at the time. Fifty years afterward, in the new millennium, those sentences have a predictive tone. Sagan wrote that the advanced extraterrestrial society would practice "self-selection on its members. The slow, otherwise inexorable genetic changes which might in one of many ways make the individuals unsuited for a technical civilization could be controlled."[20]

What did Sagan mean by "self-selection"? Imprisonment or execution of children with undesirable tendencies? Drugs to modify behavior, as in Aldous Huxley's *Brave New World*? Although Huxley wrote over eighty years ago, he was prophetic. Stimulants such as Ritalin have been forced onto "unruly" schoolchildren. In the future, adults might be required to take psychotropics. Compulsory abortion might be a future method of "self-selection." In China, forced abortion has limited population; in tomorrow's world, state-mandated abortion might complement genetic research that identifies "violent" or other undesirable genes. ET may be a pacifist through a social control that most people today find repugnant. Sagan opened a can of worms. Drake, Morrison, and Bracewell wisely remained silent.

The Debate on Extraterrestrial Life

Speculating as to whether or not extraterrestrials are benevolent assumes they exist. Their existence was debated in antiquity, and the pluralists continued the debate into the early modern era. With most pluralists, however, a deity was still necessary to explain life's origins. Only with Darwin did the secular, modern rationale for extraterrestrial life begin. Although Darwin's theory of evolution did not deal with life's origins, but with speciation, the theory has been extended to imply that life arose naturally not only on Earth but also on other worlds. Yet, strangely enough, biologists have been wary about alien life. Soon after Darwin proposed evolution, it was astronomers such as Camille Flammarion (1842–1925), Giovanni Schiaparelli, and most notably Percival Lowell who saw in Darwinist theory a non-creationist reason for the existence of extraterrestrials.[21]

In reply, Alfred Russel Wallace, cofounder with Darwin of the theory of evolution, rejected Lowell's optimistic belief in Martians. Wallace's words in his *Man's Place in the Universe* (1903) were nothing new, but Wallace spoke with gravitas. Terrestrial life resulted too much from pure chance, said Wallace; if Earth is typical, alien intelligence is unlikely.[22]

Throughout the remainder of the twentieth century, biologists echoed Wallace's skepticism, pointing out that in the span of billions of years countless freak events could have stopped or changed Earth's evolutionary progress.[23] Perhaps the key freak event occurred at the end of the Cretaceous period some 65 million years ago, when an asteroid hit Earth, killed off the ruling dinosaurs, and allowed mammals to triumph. As for humanity itself, if Mitochondrial Eve, who lived in East Africa some 200,000 years ago, had died at birth and thus been prevented from being the ancestor of every living human, if some other crucial event in evolution's chain was likewise missing—the "ifs" can be endless—would humans walk on Earth today?[24]

With the start of the SETI era, the debate was relaunched. The paleontologist George Gaylord Simpson (1902–84) noted the "curious fact" that physical scientists were more likely than biologists to accept the possibility of extraterrestrial life. Simpson pointed to an insight, first enunciated by W. D. Matthews in 1921, that biologists look for "positive evidence," whereas physical scientists have a weakness for hypotheses that cannot be "definitely disproved."[25]

Drake illustrates Matthews's insight. Drake concedes the complexity of the fossil record and the ever-changing nature of physical characteristics. These concessions aside, Drake discerns a "current" in evolution, namely, the increase of intelligence over the eons. Mammals are smarter than dinosaurs; humans, whose arrival is very recent, are smarter than other mammals.[26]

Drake's "current" harks back to the philosophical doctrine of teleology, which sees purpose in the material world. This can lead to the contention that in spite of its modern gloss the SETI narrative is simply an old idea, another instance of dressing up traditional beliefs in new words and settings, and claiming the whole thing is novel. Throughout history, humans have peopled the heavens with imaginary beings.[27]

SETI partisans of the 1960s counterattacked with two arguments in favor of extraterrestrial biology. One argument, though, is dated, for it depended on the planetary ignorance of the 1950s and early 1960s, when astronomers could still fancy primitive life on Mars and Venus. In 1961, Melvin Calvin compared Venus to Earth; each had roughly the same size, gravity, and carbon atmosphere. Calvin of course knew of the dense atmosphere covering Venus, but the unseen surface made him optimistic. In 1962, Philip Morrison commented on "the probable existence of some form of life on Mars," adding that he badly wanted Martian life because it meant Earth's life was not unique. Discovering Martian life would "transform the origin of life from a miracle to a statistic." That same year, Frank Drake wrote that he was very certain of "a very primitive, little evolved life" on Mars. As evidence, he pointed to the "dark regions that change with the seasons."[28]

Drake, Morrison, and the other SETI advocates were living in the last age of interplanetary innocence. Illusions began fading when the Mariner 4 probe of 1964 revealed a Martian surface moonlike in appearance and dotted with craters. The Mariner 9 probe of 1972 did reveal a huge volcano, Olympus Mons, raising hope for an underground life on the Red Planet or at the very least traces of extinct life.[29] To this day, though, life of any kind has yet to be found on Mars.

Martian or Venusian life would have been nice, but SETI's main argument for extraterrestrial existence during the 1960s was something else. A new view of planets had emerged in which the solar system was no longer unique, no longer the result of a near collision between a wandering star and the sun. The theories of James Jeans and Arthur Eddington, dominant in planetary thinking for much of the twentieth century, were now abandoned. The nebular hypothesis was back in favor. With the planets of the solar system no longer astronomical freaks, planets could very well exist beyond the solar system, and they might have generated living creatures.[30] Also pointing to extraterrestrial life was a new view of the sun as a "second generation" star, created from the debris of older stars that had died. Long before intelligent life evolved on Earth, it may have emerged in older worlds around these stars. Older and more advanced alien civilizations were less improbable.

If the creation of the solar system was a normal event, repeated countless times in the universe, the question naturally arises whether the emergence of life itself is common. To this question, SETI gave an affirmative answer, one that rested on updating an old metaphysical concept, the argument from analogy, beloved of the pluralists of old.

The modern version of the analogy principle is the badly named principle of mediocrity, which considers Earth "mediocre" because it is average, not because it is inferior. To avoid confusion, the mediocrity principle should be renamed the "normality" principle inasmuch as Earth's crust is normal in the cosmic scheme of things. According to spectrographic analysis, Earth's elements are found throughout the galaxy and beyond. Furthermore, normality holds true for physical laws. In Sagan's words: "The same interactions occur [everywhere in the universe], the same laws of nature govern their motions."[31] Apart from chemistry and physics, size also matters. In his 1962 book, Drake estimated that one in ten of the galaxy's stars were roughly the same size and age as the sun. Although extrasolar planets were still awaiting discovery, Drake suspected, as did most astronomers, that these planets existed and eventually would be detected.[32] In the 1990s these suspicions were confirmed.

The principle of mediocrity extends beyond dead atoms and lifeless gravity. According to the principle, the same elements that generated living matter on Earth should produce the same miracle elsewhere. On Earth, once life began, its various forms competed to survive; the same competition should take place elsewhere. Since intelligence conferred advantage in the Darwinian competition, intelligent creatures should likewise emerge in other worlds. The only unknown is the prevalence of intelligence.

Sagan did concede that extraterrestrials might not physically resemble humans. All the same, he insisted that cognitions should roughly match. All creatures, regardless of body chemistry and physical structure, live in a universe of identical physical laws. They cannot escape the effects of gravity, temperature, and other physical fundamentals, and must adjust accordingly. In short, according to Sagan, physical laws are behavioral destiny. Much human and alien thinking, when not nearly identical, must be similar. Sagan had to believe in a convergence of thought, or else contact with aliens was somewhat meaningless.[33]

Like Hoyle, Sagan envisioned extraterrestrials happily exchanging information. Once humans joined the Galactic Club, they too would acquire useful knowledge and perhaps, as a bonus, learn how to save themselves. Sagan was extrapolating from the cultural interchange of modern times in which other cultures are visited for learning and guidance; one example is the Western interest in Eastern religions and philosophy. Sagan and the SETI community believed in Earth practice writ large: extraterrestrials on various planets were receiving advice and knowledge from each other.[34]

Bracewell's Interstellar Probes

Although convinced of contact's benefits, the SETI community agreed less on the best means to reach extraterrestrials. Not everyone was sold on radio waves. Dyson had urged searching for infrared heat. The so-called Dyson spheres he envisioned would emit this heat in huge quantities. Bracewell had another search method: looking for alien probes.

Soon after Morrison and Cocconi suggested that extraterrestrials were beaming radio signals, Bracewell published his own interpretation of the alien mind. In his anthropomorphic scenario, extraterrestrials were beset by power shortages and unable to meet the ravenous power needs of interstellar transmitters. Moreover, extraterrestrials were clueless as to which star system they should direct their beacons or very strong signals. An extraterrestrial civilization near a thousand star systems needed a thousand transmitters to search thoroughly. Bracewell concluded that power shortages made extraterrestrials rule out radio beacons.

Driven by energy deficits and ignorance, extraterrestrials need another contact strategy, Bracewell reasoned. Instead of transmitting electromagnetic signals, they might "spray some number of suitable stars . . . with modest probes." Using stellar power, each probe could orbit itself in a star system, assess a planet's technology, send out signals, and wait for replies. The probe method of contact, according to

Bracewell, would require far less energy than home-based transmitters and free the receiving civilization from choosing a compatible frequency for its receivers. Bracewell's conclusion: instead of aiming radio telescopes to selected stars, astronomers should look for a probe in the solar system. If found, the probe might repeat our signals to confirm contact. Bracewell further speculated that the probes are stocked with information; its computers are ready to converse.[35]

Morrison and Drake were not impressed. Morrison noted the serious problems of erosion and maintenance and asked whether a probe could long survive in the hostile environment of space.[36] In his rebuttal to Bracewell, Drake used one of his favorite defenses of radio SETI: the vastness of space. Unless the probe did the impossible and traveled faster than the speed of light, Drake insisted, its journey would be excessively long, spanning decades if not centuries and millennia. The same physical law that made the speed of light an absolute maximum for human space travel also applies to extraterrestrials. Drake agreed with Morrison: extraterrestrials saw the futility of probes and were instead beaming radio signals.[37]

In a 1983 interview, Bracewell defended his probe theory, claiming that it had been attacked on "strikingly emotional and irrational grounds," the work of "action-oriented people" who lacked patience and wanted instant gratification instead of waiting for extraterrestrials to contact Earth.[38] To give Bracewell his due, a case can be made that waiting for a radio message or waiting for a probe makes little difference. No one knows the alien mind.

Carl Sagan

Although Bracewell was a star in the SETI world, to outsiders he was not the best-known SETI personality, nor was Morrison or Drake. That honor went to Carl Sagan, who often wrote and spoke of extraterrestrials. Sagan's fame, however, was less due to his musings on aliens than to his presentation. He had a knack for getting the public's attention, quite unlike other science professors who published in obscure journals and were known only to their peers. Even professors with salable talents who consulted for government and industry could not match Sagan's notoriety. They reached the attention of the outside world only when the *New York Times* published their obituaries.[39]

Sagan was in a different league. Instead of academic jargon, he wrote in clear English and popularized science in best-selling books. Charming and blessed with a voice that keeps attention, Sagan was a natural for television. In the 1970s he reached the wider public by appearing on Johnny Carson's *Tonight Show*. He was so effective in promoting science that he landed his own show. Sagan both produced and starred in *Cosmos*, a PBS series on the history and future of science. Televised in over one hundred countries, *Cosmos* pushed Sagan's agenda, a belief in science tempered by fears for its misuse, especially nuclear war. *Cosmos* made Sagan the best-known American scientist of the second half of the twentieth century. By the time he died in 1996, Sagan was Mr. Science to the general public.

His media gifts were only one reason for Sagan being an academic superstar. Sagan had the ego of every successful showman; he loved being the center of attention—so much so that many of his academic peers despised him. He ended up at Cornell University in 1968 because Harvard University refused him tenure, an appointment that usually recognizes peer popularity as well as scholarly achievement. To the day Sagan died, his standing with his fellow scientists never matched his fame with the public.

Besides ego, Sagan had another trait that pushed him into the spotlight; he had the missionary impulse. To Sagan, true knowledge was empirical; if the senses or their technological extensions did not perceive an alleged phenomenon, it did not exist. God was not seen; therefore, according to Sagan, God's existence was problematic—likewise with little green men from space. The masses should know this; instead, they were superstitious and ignorant, or so Sagan believed. Educating the masses was his self-appointed mission.

Strangely enough for a confirmed skeptic, Sagan flirted with ufology early in his career. In 1963 he published an article in which he seemed to accept the reality of flying saucers, claiming that extraterrestrials might have visited Earth in the past. Sagan also claimed that "billions of years of independent biological and social evolution" had made humans and extraterrestrials think very differently, making a radio conversation "very difficult."[40]

Why was Sagan negative toward radio SETI? He was impatient. At the speed of light, a radio signal from a planet only twenty light-years distant—a planet so near that by earthly standards it was like the house next door—would take twenty years to reach Earth. Yet nearly all the planets in our galaxy are farther away, some so far that their electronic signals require thousands, if not millions, of years to arrive. The SETI culture demands patience, an extreme willingness to delay gratification.

Sagan, whose passion for science fiction had seemingly affected his judgment, had no such patience. Only a conversation would allow Sagan to learn from aliens. Radio signaling clearly was not enough. Space travel was needed for face-to-face communication, assuming ET had a face.

Sagan knew, of course, that 1960s technology could not navigate the stars. The Apollo astronauts traveled some 25,000 miles an hour. At this speed, many years are required simply to leave the solar system, and it would take more than 250,000 years to reach Epsilon Eridani and Tau Ceti, the nearby star systems Drake tried to contact with Project Ozma.

Increasing the speed is not enough. At the speed of light, a spacecraft still needs four years to reach the nearest star system, and 27,000 years to reach the center of the galaxy. Although the speed of light is fast, clearly it is not fast enough. Nevertheless, it is an absolute barrier.

If Earthlings of the twentieth century could not reach distant worlds, direct contact required that extraterrestrials come to Earth. Sagan suggested the means for this voyage: the ramjet, which the physicist Robert Bussard (1928–2007) proposed in 1960. By using hydrogen in space as a fuel, the nearly science-fiction ramjet called to mind the explorers of old who lived off the land. Sagan conceded that a vessel feeding itself from the space medium lay a century in Earth's future, but advanced aliens might be using it or something similar. Sagan was pitching the idea of extraterrestrials having mastered rapid space travel both in ways we can imagine and in those we cannot. The subtext was that extraterrestrials might arrive. Sagan's suspicion (if not a conviction) of an extraterrestrial landing, perhaps as recently as historical times, complemented this reasoning. Sagan wondered whether an extraterrestrial visit led to ancient religions whose gods live in the sky. Sagan carefully noted that hard evidence for this visit was lacking, yet his words could be ripped from context and seen as endorsing ufology.[41]

Sagan was willing to discuss ufology. Unlike other scientists who feared ridicule and dismissed alien visitations, he was not afraid of taboos. He may have wanted to discuss ufology because he felt a kinship. Like the ufologists, he passionately believed in extraterrestrials, but with one big difference. Sagan denied that extraterrestrials were routinely visiting Earth, although he suspected they might have paid a visit or two in the past. Yet, he demanded proof, scientifically verifiable physical traces from an alien visit, past or present. Sagan was fond of saying that extraordinary claims require extraordinary evidence, and he believed the public overly credulous when the popular media publicized tales of extraterrestrial visits.

In 1961, when his thirtieth birthday was three years in the future, Sagan began his career of educating the public when he volunteered to be a prosecution witness in a fraud trial. A sixty-year-old Nebraska salesman had sold stock in a quartz mine whose rocks allegedly cured cancer. According to the defendant, spacemen from Saturn had told him of the mine. Sagan testified that Saturn's environment was totally unable to support humanoid life. The defense counsel countered by asking Sagan if scientists did not once believe in a flat Earth. Suddenly, as Sagan later wrote, a fraud trial had become a defense of modern science, as he had to demonstrate why scientists could accurately describe the temperature, gravity, and atmosphere of an unvisited planet millions of miles from Earth. Sagan worried that the clever defense lawyer was impressing the jurors who sat impassively. Sagan underestimated the jurors' intelligence or maybe his own charisma. The defendant was convicted of fraud.[42]

The trial was a footnote in Sagan's rich biography. In the 1960s, Sagan taught at Harvard, consulted for the Pentagon and NASA, and published in both scientific and popular journals. Although busy, he never lost the urge to enlighten the masses. When Stanley Kubrick was filming his 1968 movie 2001: A Space Odyssey, Sagan made an important contribution. Sagan would have called it a correction. In the movie, extraterrestrials guide human evolution and progress, but Kubrick was uncertain of how to depict them. Sagan insisted that the appearance of the aliens was a moot point. No one knows their true appearance, for their evolutionary history very likely differs from ours. To make the movie accurate and avoid deceiving the public with questionable alien images, Sagan advised Kubrick against showing them. Kubrick accepted Sagan's advice.[43] This ruined the movie for many moviegoers. Every science-fiction movie ends with the revelatory scene, when the monster or alien is shown. Audiences expect this scene, and many moviegoers left Kubrick's movie feeling cheated. To this day, *2001* remains controversial, a masterpiece to some and frustratingly obscure to others.

Always willing to dispel ignorance, in 1968, along with astronomer Thornton Page (1913–96), Sagan proposed that the American Association for the Advancement of Science (AAAS) hold a symposium on UFOs. In an exchange of letters, Sagan and Donald Menzel (1901–76), a Harvard astronomer well known for debunking UFO claims, debated the wisdom of staging the symposium. Menzel feared that the AAAS was validating ufology by discussing UFOs. Sagan disagreed, believing that "a confrontation of alternative views" served to expose "questionable logic or unjustified conclusions." Besides, added Sagan, the symposium would publicize the "scientific traditions of fair play and free inquiry." Menzel was dubious, but nonetheless yielded.[44] Along with Morrison, Drake, and Walter Sullivan of the *New York Times*, Menzel attended the symposium, which was held in Boston on December 26 and 27, 1969.

The AAAS symposium happened to follow the publication of the Condon Report (1968), a government-sponsored investigation of UFOs, which rejected their alien origin and recommended the suspension of Project Blue Book, the Air Force program launched in 1952 to study UFO phenomena. Backed by the Condon Report, the AAAS symposium was, not surprisingly, skeptical of UFO allegations. As could be expected, ufologists dismissed the Condon Report and claimed that the fix was in, as they have done ever since.[45]

Sagan failed to realize that ufologists were not ignorant, that they rather possessed a different extraterrestrial narrative. Perhaps, deep down, Sagan was attempting to reach out to ufologists. Although Sagan never claimed outreach was his intention, if ufologists saw their errors, they might support SETI.

SETI did need help, having hardly progressed during the 1960s. Neither private nor public money had materialized, and without funding for a proper search, SETI, at least in the United States, was destined to remain a discussion topic for university professors and elite members of the public. If ufologists realized that George Adamski and his ilk were frauds, they might support SETI, send letters to Congress, and create a public clamor that might shake SETI money from NASA. In the real world, though, the chances of ufologists following the lead of egghead professors were slim. If Sagan was looking for allies, he was looking in the wrong place. Ufology did not result from ignorance; it resulted from an anti-elitist streak in American life.

In the 1960s, Sagan was best at writing. His great contribution to SETI was his book *Intelligent Life in the Universe*, coauthored with the Russian astrophysicist I. S. Shklovskii (1916–85). The book had originated with Shklovskii, who published the Russian edition in 1962. Sagan revised and extended the Russian edition, doubling its size with background on biology and history. The result was still a heavy tome, but it was far more reader-friendly than the original Russian version. Published in 1966, the American version went through fourteen editions by 1975.[46] It converted more than one reader to SETI. American SETI would have a better decade in the 1970s.

6

Soviet SETI

ET Is a Communist

In the early 1960s, the Soviet Union appeared to be winning the space race. On April 13, 1961, the Red superpower sent the first man into space, Yuri Gagarin (1934–68). Less than a year later, it sent the first woman into space. On March 18, 1965, another Russian walked in space. Supporting these Russian heroes was some very impressive equipment: Soviet rockets and space capsules far larger than the American ones. How could the Soviet Union top these space achievements? One way was to receive the first extraterrestrial message.[1]

Russian Cosmism

Soviet and American SETI had different origins. Whereas American SETI had its distant causes in the pluralist tradition and its nearer causes in Cold War anxieties, Soviet SETI had its roots in Communist ideology and in Russian cosmism, an intellectual movement that originated in the late nineteenth century and has persisted to the present. Cosmism can be described as a unique blend of science and magic, tradition and futurism, religion and materialism. One theme, though, runs through all the various strands of cosmism: the human race has as active role in its evolution. Humanity will escape its earthly prison, roam the cosmos, and settle in the worlds of space.[2]

A leading figure in cosmism was Nikolai Fedorov (1828–1903), a Moscow librarian and autodidact. In his spare time, the ascetic Fedorov read nearly every book in his care and presided over a discussion group that Dostoyevsky and Tolstoy frequented. Fedorov believed all matter was alive, including sand and rocks, which differed from obvious forms of life only by lacking consciousness. As the highest form of consciousness, the human race had the duty to regulate the world and the universe. The Soviet Union's obsession with the conquest of nature as seen in the huge dams and irrigation projects of the Five-Year Plans owes something to Fedorov and cosmism.[3]

For Fedorov, the greatest human problem was death. Only its abolition would solve other problems, whether they be political, social, or economic. Once humanity's unlimited potential was channeled toward ending death, not only the living and their descendants would enjoy immortality but also the dead through resurrection. Unfortunately, the fulfillment of this dream, the conquest of death, implied an Earth running out of room, forcing humanity to settle on other worlds. The simple logic of Fedorov's philosophy left no alternative.[4]

Fedorov's protégé, Konstantin Tsiolkovsky (1857–1935), showed the way to the stars. Born poor and nearly deaf, Tsiolkovsky never had the chance to attend university. Coming to Moscow, he impressed Fedorov, who supervised his informal education and later placed him in rural Kaluga (southwest of Moscow) where Tsiolkovsky spent his life teaching mathematics.[5] Here, Tsiolkovsky did basic work in rocket science and space flight theory. He wrote equations that calculate how much fuel a rocket needs to escape Earth's gravity; he tested the effects of rapid acceleration on organisms; he designed gyroscopes and conceived multistage rockets.[6] For these and other contributions as well as his humble origins that proclaimed the new Soviet man, Tsiolkovsky was celebrated in the Stalinist era. His cutting-edge research dovetailed well with the Soviet insistence that Communism was the future.[7] His science-fiction pieces, both long and short, inspired many future Russian rocket scientists, including the team that launched Sputnik I. His readers also included Yuri Gagarin.[8]

Less known is Tsiolkovsky's belief in extraterrestrials. The Communists ignored this side to Tsiolkovsky, deeming it irrelevant to his scientific work, although Tsiolkovsky had intimately tied his science to his philosophy of life. He believed higher forms of life everywhere were destined to leave their worlds and expand, and that the human race would also one day reach the stars.[9]

Tsiolkovsky derived his view of the universe from monism, a philosophy that sees spirit and matter as one, denies diversity, and reduces everything to sameness. The ultimate reality were atoms of ether, eternal and sentient, making the universe alive. Like the atoms of the Greek atomists, Tsiolkovsky's atoms wandered and combined to form higher stages of life. A universe fundamentally the same throughout meant higher life was everywhere, both on Earth and on the many worlds of the stars. Humanity belonged to the universe as much as it belonged to Earth.[10]

Like pluralists of old, Tsiolkovsky argued from analogy. His 1933 essay "The Planets Are Occupied by Living Beings" claimed that life on Earth pointed to life on other worlds. Like Lowell, he equated age with progress.[11] Tsiolkovsky was convinced that older extraterrestrial races were more developed and were communicating with humans, a one-way activity of which humans are oblivious. Geniuses may owe their inspiration to these forces beyond their awareness. Tsiolkovsky also anticipated future skeptics who have asked why advanced extraterrestrials have not landed on Earth. His answer anticipated that of later SETI apologists: the aliens simply had not yet arrived. After all, said Tsiolkovsky, before the eighteenth century, Australian aborigines had not seen a European.[12] As for why extraterrestrials have stayed hidden, Tsiolkovsky reasoned that extraterrestrials have judged humanity primitive, violent, and best left wallowing in its ignorant pride.

The New World of the Soviet Union

After the 1917 Revolution, enthusiastic Communists believed that a new era of science, arts, and learning was dawning. However, in the real world, which even revolutionaries had to live in, matters were not so simple. Although traditional intellectuals could be silenced, fired from teaching posts and refused publication, if not shot or exiled, old ideas have a way of enduring. Pagan folkways that found a home in

Christianity come to mind. Cosmism probably had its roots in a pantheism predating Orthodox Christianity, and as a strain in intellectual life it survived after 1917.

The Soviet Union saw itself as the vanguard of modern culture, thus opening the door to novelties. Soviet culture flourished in the 1920s when creative minds unleashed themselves. Retrenchment came in the 1930s, when the philistine hand of dictator Joseph Stalin, who ruled the Soviet Union from the 1920s to his death in 1953, slapped down the movie directors, poets, and artists who did not share his pedestrian tastes.

In the Stalinist era, the search for novelty did not necessarily disappear. It merely mutated, driven by the need to demonstrate Soviet superiority over capitalist countries, a tough task since the Soviet Union boasted few achievements. The Soviet Union was a cultural wasteland, and its economy, apart from the rapid increase in capital goods, could not compare with the West's. The good life of plentiful food and the consumer baubles of the twentieth century was embarrassingly absent in the so-called workers' paradise. The result was a susceptibility to quick fixes, "unorthodox by bourgeois standards, in which difficulties might be solved, or decisive savings made" for the Soviet economy.[13] The quick-fix mentality explains the career of the great charlatan of Soviet biology, Trofim Lysenko (1898–1976), who rejected standard Mendelian genetics with its time-consuming procedures and in its place proposed to improve plant stock by changing the physical environment. Contacting extraterrestrials and acquiring their knowledge was another quick fix. The enhanced prestige of the Soviet Union was an added bonus.

Moreover, Communist dogma hinted at the existence of extraterrestrials. Vladimir Lenin (1870–1924), founder and first leader of the Soviet Union, subscribed to a thoroughgoing materialism in which the development of matter inevitably generated life and intelligence. Logic required that the same evolutionary forces were present on other worlds. Extraterrestrials could validate Communist ideology, which as all good Communists knew was scientific and universally valid. Aliens would give the ultimate proof of the logic of atheism and of a Communist society. With luck, aliens might share military secrets with the people who first contacted them. It was no accident that the Soviet military had sponsored research in "such subjects as ESP and time travel, hoping for any kind of counter to the West's economic and technological advantages."[14]

Although Soviet and American SETI used the same detection method of checking whether radio signals from space were natural or artificial, the SETI cultures of the two nations decidedly differed. American SETI was distinctly American, a grassroots initiative of scientists who believed in radio contact with extraterrestrials and lobbied their colleagues and the public. Drake and Morrison resembled entrepreneurs who create a new product and convince the public of its necessity. By contrast, Soviet SETI took place for the same reason the Soviet Union boasted excellent sports teams and ballet companies. The Soviet bureaucracy had decided that SETI, like Olympic medals, promoted the Soviet cause. Presumably, Soviet bureaucrats had needed convincing, but Soviet SETI advocates had lobbied in private. Once the Soviet powers-that-be decided to take a chance on SETI, Soviet SETI scientists, unlike their American counterparts, did not need to defend their legitimacy in public. In spirit, Soviet SETI was as distinctly Soviet as the pervasive Lenin cult and the ubiquitous secret police, the KGB.

Shklovskii and Kardashev

Soviet SETI is intimately associated with I. S. Shklovskii, the famed theoretical physicist and radio astronomer. Remembered mainly for having founded Soviet SETI, Shklovskii had achieved in his pre-SETI days a reputation at Moscow State University's Shternberg Astronomical Institute, where he taught from 1938 to his death. For his discoveries on cosmic rays and the sun's radio waves, he won the Lenin Prize in 1960. Although a mainstream astrophysicist, Shklovskii could be idiosyncratic. In 1959 he suggested that Phobos, the larger of the two moons of Mars, was hollow and possibly of artificial origin.

In 1962, Shklovskii published his seminal work on SETI, *Universe, Life, Intelligence*, which Carl Sagan expanded and renamed *Intelligent Life in the Universe*. Sagan distinguished his contributions from Shklovskii's so that the reader knew which author was writing. In the original Russian edition, Shklovskii had referred to the superiority of Communism and of its necessity for world peace. Sagan noted these views were Shklovskii's, not his own. Chances are that the views were not Shklovskii's either. Soviet censors habitually vetted manuscripts before publication and were not averse to adding and deleting. Shklovskii's later criticism of the Soviet atomic bomb program makes his enthusiasm for the Soviet system suspect. There had to be a reason why for a long time he was not permitted to leave the Soviet Union. In the preface to their joint book, Sagan admitted to never meeting Shklovskii because, as Sagan delicately put it, "he does not travel out of the Soviet Union."[15]

Although the theoretician of Soviet SETI, Shklovskii never did his own searches. He inspired others, especially his star pupil, Nikolai Kardashev (b. 1932), also a professor at Moscow State University. Kardashev had dreamed of an astronomy career from the age of five, when his mother took him to the Moscow planetarium. When he later read the writings of Flammarion and Schiaparelli on Martian life, his enthusiasm grew. After the publication of the Morrison/Cocconi paper, he seriously thought of contact. About a year later, a special unit was formed in Moscow to explore the possibility. In 1963, Kardashev conducted the first Soviet SETI search when he examined Quasar CTA-102.[16]

Kardashev is famous for his ingenious system of using energy consumption to classify extraterrestrial civilizations. A civilization exploiting all the energy resources of its planet and nothing more Kardashev labeled "Type I," and it stands at 1 on the Kardashev scale. (Today's Earth stands at 0.72 on the Kardashev scale.) An extraterrestrial civilization using the entire energy of its star system he labeled a "Type II." More advanced is the "Type III" supercivilization, which harnesses the energy of all the stars in its galaxy.[17] Defining progress in terms of energy consumption is a quaint notion in our twenty-first century, which emphasizes efficiency, but quite the norm for the energy-guzzling twentieth century, and especially for the Soviet Union. Lenin had famously said that Communism was Soviet might plus electrification. Lenin believed the Soviet Union was doomed without electricity for industry and agriculture, hence the huge dams and hydroelectric systems of the Five-Year Plans. Living in a power-obsessed society, Kardashev probably never conceived of advanced extraterrestrials living with less. His extraterrestrials did not belong to the intergalactic Green Party.

Soviet Astronomers Meet at Byurakan

On May 20–23, 1964, at Kardashev's suggestion, Soviet astronomers met at the Byurakan Astrophysical Observatory in Soviet Armenia. The director of the observatory, V. A. Ambartsumyan (1908–96), laid out the premise underlying the conference: "Life and civilizations exist on a multitude of celestial bodies."

Unlike American SETI discussions, there was no defense of SETI at Byurakan. A political decision had been made to develop a search program for extraterrestrials. The aim of the conference (again in Ambartsumyan's words) was "to obtain rational technical and linguistic solutions for the problem of communication with extraterrestrial civilizations which are much more advanced than the Earth civilization."[18] The key word here was "communication." It implied a dialogue with extraterrestrials. In fact, the term "SETI" was not yet invented. Until the Americans adopted this term in the 1970s, the movement to contact extraterrestrials was referred to as CETI, or Communication with Extraterrestrial Intelligence.

In his opening remarks, Shklovskii hoped to see a key question answered in five years: whether there was life beyond Earth. He was sure of a positive answer. After all, in 1964 many astronomers still believed that Mars contained a primitive life. Shklovskii could not resist taking a shot at the Soviet establishment. In searching for extraterrestrial life, Shklovskii dismissed "materialistic philosophy" as a guide. "Only scientific arguments, primarily experimental work and observations," could supply the answer.

Gradually, Shklovskii warmed to his main argument: Soviet SETI should focus on Type II and Type III civilizations. Unlike Morrison and Drake, he doubted that civilizations on Earth's level were transmitting signals. The Russian did agree with his Western counterparts on one point, though: intelligent life was moving toward greater cognition and knowledge. The existence of extraterrestrial races older than humanity and advanced due to their head start were a given for Shklovskii and also for American SETI.

Extrapolating from Earth's history, Shklovskii predicted rapid increases in energy consumption. In less than three hundred years, he believed, heat from energy production might cause "a very acute problem." Like Kardashev, Shklovskii did not foresee technologies that consume less power and radiate less heat. Shklovskii feared a frying of Earth unless energy production moves into outer space. Once in space, the human race will "conquer and transform the solar system" within ten thousand years. Shklovskii believed that a number of interstellar civilizations had already reached this level, having continued to expand until they finally colonized their entire galaxies. All this took place within a few tens of millions of years and maybe fewer, since "expanding super civilizations develop qualitatively, as well as quantitatively." Shklovskii's conclusion was simple: if advanced extraterrestrials exist, they have revolutionized their worlds, and their past reveals Earth's future. We see our tomorrow neither in tea leaves nor in algorithms, but in the stars.

Shklovskii asked why advanced extraterrestrials would contact a backward humanity. He answered his own question by citing recent history. The New York World's Fair of 1939 buried a time capsule containing artifacts and microfilms so that people in the year 6939 can comprehend the twentieth century. Shklovskii said: "This undertaking in principle is not different from communication with alien civilizations. The only difference is the cost." Shklovskii was implying that aliens had the human trait of pride. They transmitted their version of the encyclopedia in order to be remembered and appreciated.

Like the Americans, Shklovskii also believed in radio, which he defined as "the most economic and informative among all the conceivable means of communication." Shklovskii failed to see that radio best applied to Type I civilizations. A Type III civilization so advanced that it controlled the energy resources of its galaxy must have developed a superior means of communication.

In spite of official backing, Soviet SETI's resources were limited. In a 1981 interview, Kardashev mentioned opposition to SETI.[19] One reason might have been a higher priority for physics in the energy-obsessed Soviet Union. At Byurakan, Shklovskii complained that fusion researchers enjoyed ample resources yet were excused from producing quick results. By contrast, Shklovskii complained, the importance of finding extraterrestrial civilizations could "hardly be overestimated." He was hinting that given a few more rubles, he could find ET and enhance the Soviet Union far more than if fusion researchers created cheap energy. It can be assumed that Shklovskii did not get the rubles. In April 1965 the *New York Times* reported that Soviet astronomers were demanding more funding at the expense of physics.

A Soviet scientist requesting more funding for his research was one thing, but denying Marxist optimism was something else. Citing earthly models, Shklovskii speculated on the death or degeneration of extraterrestrial civilizations. Like his American counterparts, Shklovskii was prone to pessimism. (No wonder he got along well with Sagan!) However, Western-style *Weltschmerz* was dicey in the officially upbeat Soviet Union. One scientist at the conference rejected Shklovskii's short timeline of ten thousand years for an extraterrestrial civilization, asking why it could not last a billion years.

D. Ya. Martynov of the Shternberg Astronomical Institute gave the discussion the "correct" perspective. He reminded his audience of the Marxist line: "Human society evolves according to dialectical laws," and the struggle of antagonistic forces should not breed pessimism. "Once a classless society" is established, he said, humanity could expect "a comparatively calm stage with more rational and economic exploitation of resources." The implication was obvious. Marxist philosophy guarantees a happy outcome to human history. After worldwide revolution, the scourges of war, pollution, and poverty will end. Since Marxist laws of society are universal, they also apply to aliens. Extraterrestrial civilizations can be immortal. This line of reasoning was good for Kardashev, for it reinforced his theory of supercivilizations, if not being the root cause.

Whether Ya. Martynov and the others believed this Marxist millennialism is irrelevant. They had to express it. Although Stalin had been dead for eleven years, his system had left its lingering effects. Only a fool did not realize that at least one scientist at Byurakan was spying for the KGB. Surveillance was standard in the Soviet Union, and its citizens factored this fact into their daily routines. Shklovskii's pessimism was veering toward dissent, and the other scientists at Byurakan had to cover themselves.

When Kardashev spoke he echoed Shklovskii's obsession with energy, predicting that in 5,800 years humanity would need the energy of 100 billion stars (the estimated size of the Milky Way galaxy in the 1960s). Some older civilizations had already reached this Type III status, Kardashev reasoned. Although these supercivilizations had not been detected, Kardashev added that no data ruled out their presence.

Like American SETI, Soviet SETI had the challenge of identifying alien signals. Kardashev's strategy assumed that supercivilizations were unaware of other civilizations. This ignorance was forcing curious extraterrestrials to transmit in all directions, sending wide-band signals for easy detection. This scattergun approach demanded a huge energy supply, which only supercivilizations could provide.

Kardashev's reductive reasoning failed to convince all of his listeners. Ambartsumyan questioned the virtues of wide-band signals, noting that cosmic noise would have less effect on narrowband signals. Another speaker asked how Type III civilizations could control billions of star systems and yet remain undetected. When Ya. Martynov asked why Type III civilizations bothered to transmit huge volumes of data, P. M. Geruni answered by distinguishing "healthy" from "unhealthy" civilizations. The healthy ones were long-lived and eager to gain new information by sharing their own, whereas the unhealthy ones were self-centered. In a burst of wishful thinking often seen in American SETI, Geruni said: "Transmission of information is properly considered a duty of any more or less advanced civilization."

In his defense, Kardashev resorted to Marxist dogma. A belief in progress, he said, implied a belief in supercivilizations. Civilizations had to advance, or else "some fundamental restriction" was blocking evolution. Kardashev insisted that searching for supercivilizations should come first, for finding them was "technically simpler." Despite their distance, supercivilizations would be sending continuous, strong signals, which less-sophisticated technology could catch.

The conference ended with resolutions. The first resolution validated

SETI as a legitimate field of science. Instead of specifying a favorite radio frequency, as Morrison and Cocconi had done, the conference recommended a wide range of frequencies, from 1 GHz to 100 GHz. On the crucial issue of whether to look for nearby Type I or distant supercivilizations, the conference waffled, endorsing both strategies; the final decision was deferred to committees that would research the necessary technology. Finally, it was decided to meet again the following year, but this meeting was not held.

Optimism had marked the conference. In contrast to American SETI, which saw contact as a possible means to cure earthly ills, Soviet SETI viewed contact as a logical outgrowth of human progress, the vanguard being of course the Soviet Union.

In this climate of anticipating success, Soviet SETI's false alarm of 1965 is understandable. Soviet observers detected a radio emission, which Kardashev had previously postulated as the beacon of an advanced extraterrestrial civilization. The Soviets could not contain themselves. Instead of taking a cue from Frank Drake and his caution during Project Ozma when he detected a possible signal (a wise move since the signal was terrestrial), the Soviet astronomers talked to TASS, the Soviet news service. The *New York Times* got wind of the story, and news of another Soviet scientific triumph briefly regaled the world. The truth was embarrassing. ET was not calling. Soviet astronomers had detected a quasar, as Caltech astronomers quickly told them. Discovered in 1963, quasars are distant galaxies that give off enormous amounts of energy.[20]

Despite the fervor at Byurakan, Soviet scientists did not quickly follow up with a SETI search. Perhaps the fallout from the quasar fiasco had embarrassed Soviet SETI, or maybe it suffered the same problem as its American counterpart: gaining access to radio telescopes. In the United States, conventional radio astronomers feared SETI as a rival in a zero-sum game. If a SETI researcher gained radio telescope time, a conventional astronomer lost out. The same may have held in the Soviet Union.

The next Soviet SETI search did not take place until 1968, when Vasevolod Troitskii (1913–96) of the Institute for Radiophysics of Gorki State University conducted sixty-five observations of twelve nearby "G" (sun-like) stars. Unlike Project Ozma's one channel, Troitskii tuned in on twenty-five channels.[21] Troitskii also did observations from September to November 1970. He later admitted that most Soviet astronomers were indifferent toward SETI.[22] The Soviet science establishment evidently regarded SETI as a gamble. SETI might succeed and bring glory to the Soviet Union, but the search was not worth massive resources.

Sagan and Kardashev Plan the 1971 Byurakan II Conference

Soviet astronomers again met in Byurakan on September 5–7, 1971. Byurakan II was international, with fifty-four participants, mostly from the United States and the Soviet Union, as well as a few from the United Kingdom, Hungary, Czechoslovakia, and Canada.

Since about 1967, Sagan had thought of having the world's SETI scientists meet. Science gains from sharing, and both Soviet and American SETI researchers were curious about their counterparts. The meeting had to take place in the Soviet Union, however, because during the Cold War, as Shklovskii's career attests, Soviet scientists could not easily travel abroad. While collaborating on his joint book with Sagan, Shklovskii wrote the American that an extraterrestrial landing was more likely than the two men meeting. Soviet delegations to the West invariably included handlers to prevent defections.[23]

The idea of holding a SETI conference that included the Americans appealed to Kardashev. However, he wanted to exclude "windbags," his term for philosophers, humanists, and social scientists.[24] Whether Kardashev sought to avoid politically sensitive topics by concentrating on technical issues or was expressing the contempt that some natural scientists have for the humanities and the social sciences is not known. Kardashev did not get his way. Sagan insisted on a multidisciplinary group; anthropologists, psychologists, and even a historian attended.

Sagan believed that the "windbags" would spice up the conference. As opposed to astronomers, they might provide novel perspectives. The "windbags" he invited revealed his conception of an adequate SETI discussion, which had to transcend the narrow world of astronomy and encompass other natural sciences, and even the social sciences.[25] For Sagan, a SETI discussion was also a discussion of the state of humanity.

The Byurakan II Participants

Among the participants were the Canadian anthropologist Richard Lee (b. 1937), who had lived with the Bushmen of the Kalahari and spoke their language. He was chosen to contribute insights on how extraterrestrials might perceive a backward humanity. Kent Flannery (b. 1934), another anthropologist, viewed culture as being in constant flux, implying a rejection of tradition. J. R. Platt (1918–92) wore two hats: professor of physics at the University of Michigan, he was also associated with its Mental Health Research Institute. William McNeill (1917–2016) edited the prestigious *Journal of Modern History*. An expert on world history, McNeill bucked the fashion of professional historians specializing in narrow topics. He would present the big picture.

The Western scientists included four veterans of the Green Bank Conference: Sagan, Drake, Morrison, and Oliver. Also present were Francis Crick (1916–2004), Nobel laureate in medicine (1962) for having co-discovered the structure of DNA; David Hubel (1926–2013), a future Nobel Prize winner in medicine for his work on visual perception; Marvin Minsky (1927–2015), an MIT electrical engineering professor and pioneer in artificial intelligence; Freeman Dyson of "Dyson spheres" fame; and Charles Townes (1915–2015), who had won the Nobel Prize in physics for inventing the laser. Townes had the added distinction of having suggested an alternative SETI strategy: looking for light pulses instead of radio signals. When Townes published this in 1961, he was ahead of his time; optical SETI emerged only near the century's end.

The Soviet delegation included veterans of the first Byurakan conference: Shklovskii, Kardashev, Troitskii, Ambartsumyan, and others. In all, thirty Soviet scientists attended.

Prior to the conference, the organizing committee had distributed a questionnaire to the Soviet scientists. One question was: "Is there a civilization outside Earth?" As expected of a SETI conference, the six Soviet scientists who responded answered in the affirmative. Their enthusiasm, though, varied. Kardashev believed in a universe of "many civilizations." Shklovskii was the least enthusiastic, responding that

Earth might be the only civilization "in the observable part of the universe." Shklovskii again stood out when responding to the question about the future of extraterrestrial studies. The other scientists gave mostly technical answers, although Troitskii did say that the life span of a civilization had to be considered. Shklovskii instead cited philosophical, biological, and futurological issues.[26] This was normative for American SETI but risky for Soviet scientists, who prudently avoided political dangers by maintaining a focus on technology.

Culture Shock for the Americans

Visiting the Soviet Union was an adventure, and the reason was neither the weather nor the food. The Americans were honest scientists seeking to expand their horizons. This meant little in the spy-obsessed world they were entering. When writing of Byurakan II, Sagan overlooked certain incidents, bagatelles in his greater worldview, but Drake mentions them.

Drake wrote two accounts. The first, from 1976, focused on the challenges of leaving the Soviet Union: the furtive luggage searches, the missed flight, the chaos and inefficiency at the airport, and the bribes. To top it all off, when the Americans reached Moscow, the hotel had paired wives and unrelated men—perhaps a statement on advanced gender relations but more likely Soviet inefficiency. Quickly, the Americans had to reshuffle the hotel's room assignments.[27] They definitely were not in Kansas.

In his history of SETI written in 1992, right after the fall of the Soviet Union, Drake was less discreet. He recalled Soviet customs at the Moscow airport, aghast when they saw the stenotype machine in the American baggage, fearing the strange contraption was a copier. One of the Americans had to be a spy, bent on copying god knows what. Although the stenotype survived Moscow customs, the incident was only the beginning. In Drake's words, the "intense surveillance of Soviet scientists and their visitors by the KGB" marked the conference.[28]

The Conference Gets Under Way

Since the Americans could not speak Russian, the Soviet organizers thoughtfully provided a translator, Boris Belitsky, science editor for Radio Moscow. Belitsky spoke excellent American-accented English, so good that he corrected grammatical mistakes. Yet many of the Russians spoke in English, perhaps eager to demonstrate their knowledge of the language of science.

Although the Americans trumped the Russians on language, they took a back seat on SETI. The only American search had been Project Ozma, which by now was old news. On the other hand, the Soviet Union was at the SETI vanguard. When Kardashev and Troitskii described their SETI work, Drake noticed the envious Americans.

The first session covered the percentage of the universe's stars blessed with planets. Vassili I. Moroz (1931–2004) of Moscow State University, whose specialty was the planets of the solar system, noted the paradox of astronomy. Although the planets are near and the stars distant, we know less about the formation of the planets, which took place billions of years ago, than about the stars' formation, which can be observed in different stages of their evolution.

In the next session, the discussion took a turn the Soviets may not have wanted. Instead of focusing on astronomy and technology, Sagan and Crick sharply debated the likelihood of extraterrestrial life, replaying the perennial dispute between skeptical biology and optimistic astronomy. SETI astronomers must believe in many intelligent extraterrestrial races, or else they are wasting their time. On the other hand, extraterrestrials are irrelevant for earth-bound biologists. They easily see contingency, all the reasons why life might have failed on Earth, all the lucky accidents that allowed it, and why it might be improbable elsewhere.

Crick used an analogy based on playing cards. If a card game has been played only once, Crick said, and if the player knows nothing of other games and players, the player cannot assume that the next hand will replicate the previous one. Many games must be played before one can calculate odds for the next hand. The emergence of life on Earth is like one deal of the cards; one cannot assume a repetition because it once happened.

Sagan replied that he was playing a different card game. In his game, no one "sequence of cards . . . wins." Sagan added that the origins of life has many paths, and "one of them has been taken on a suitable planet." Whereas Crick was playing five-card stud poker, in which only a royal flush wins, Sagan was playing poker with deuces and threes wild, in which many hands can win.

Crick was not finished. He played the spoilsport with David Hubel, who spoke on the evolution of intelligence. Beginning with current knowledge of cells and the nervous system, Hubel discussed the process of learning. Hubel was optimistic about f_i, the variable in the Drake Equation that assesses the fraction of planets that have intelligent beings. Crick was not impressed, insisting that the nervous system is poorly understood and that the origin of intelligence is very much a gray area.

By now, the Soviet delegation was mostly silent. After the first session, when the Soviets discussed the evolution of the solar system, the conference had taken a turn occasionally seen in American SETI discourse: justifying the search. The Soviets publicly avoided this topic, and the "windbags" whom Kardashev wanted to exclude had yet to speak.

Anthropologists Richard Lee and Kent Flannery discussed technological civilizations. Lee covered the four billion years from life's origins to the emergence of intelligence (humanity) and his conclusion was downbeat. He claimed the human race peaked about ten thousand years ago in its hunter-gatherer stage before it invented agriculture and civilization, the causes of pollution and possible self-annihilation.

Flannery spoke next. His first words were, "It is my job now to explain how man fell from this Garden of Eden, which Doctor Lee has described, into his present disastrous condition." Lee and Flannery shared the pessimism common to many SETI astronomers. Whether the two anthropologists realized it or not, their pessimism meant a low value to the L variable of the Drake Equation, the lifetime of a civilization that communicates to other worlds.

J. R. Platt of the Michigan Mental Health Center spoke next. Echoing Lee and Flannery's sour view of the future, Platt predicted that "half" of Earth's "critical minerals" would be exhausted within twenty years. After his geology lecture, Platt expressed his hope that aliens were biologically programmed to be cooperative and rational. If extraterrestrials resemble humanity, Platt warned, they might destroy themselves, or lose interest in progress, or suffer genetic degeneration or other traumas. Such *Weltschmerz* was bad news for SETI.

The key point of the conference was N in the Drake Equation: the number of detectable extraterrestrial civilizations in the galaxy. Sagan was hoping "a million" existed, yet he feared they had destroyed themselves. Reading newspapers, averred Sagan, was depressing. If humans were typical, older civilizations were already gone, and humanity was fated to follow.

The Soviets held no brief for pessimism. As at the first Byurakan conference Shklovskii had concentrated on supercivilizations, who, he hoped, were sending beacons to Earth. Shklovskii conceded, however, that extraterrestrials, whether on Earth's level or very advanced, could be isolationists, like the Chinese of the Middle Kingdom, indifferent to distant beings. This isolation meant that SETI should focus on supercivilizations because their enormous energy use produced strong heat emissions that were easier to detect.

Unlike the pessimistic Americans, the Russians tended to use abstract language, avoiding concrete terms such as "pollution" and "overpopulation." Instead of discussing humanity's problems, they discussed solutions. Computers were the key for Kardashev. He defined civilizations as systems to "receive and process information." The more information at their disposal, the more advanced they were. He strongly implied that expanded knowledge had allowed extraterrestrials to survive and that data crunching would likewise save humanity. G. M. Idlis was another optimistic Soviet scientist. Idlis asserted that only knowledge solves "problems" and, having doubled "every ten or twelve years" since Newton, that the expansion of knowledge will continue.

Charles Townes articulated a very human desire: the personal satisfaction of a two-way conversation with extraterrestrials. This meant searching for civilizations within ten light-years. Extraterrestrials living in deep space, thousands of light-years away, will not do. The thousands of years required for their response to reach Earth spells a very delayed gratification, to say the least. Sagan's reply to Townes was not comforting. Either many technical civilizations exist or only a few, Sagan said. If a few, it is unlikely they are bunched up in Earth's immediate neighborhood. For Townes's conversation to occur, life would have to permeate the universe so that at least a few extraterrestrial civilizations are statistically close enough.

When Freeman Dyson spoke, he lived up to his maverick reputation. Unlike Crick, who asked hard questions but did not reject SETI, Dyson wanted to pull the plug. Dyson advised astronomers to forgo SETI and do mainstream research. While reaping "a harvest of important scientific discoveries," they might find extraterrestrials as a bonus.

Kardashev ignored Dyson's call for mainstream research by stating that scientists had to consider things they "know nothing about," such as an alternate physics, in which time travel is feasible. Sagan agreed. The possibility that advanced extraterrestrials were using unknown laws of physics to communicate had to be considered, he said. In addition, these societies had to bridge the technological gap with a backward Earth and make due allowances. Marvin Minsky explained the extent of these allowances. To decipher the radio signals of 1971, a radio receiver of the 1940s would struggle. Whether Sagan and Minsky realized it or not, they were revising the Drake Equation, adding a new variable for advanced extraterrestrials willing to dumb down for humans.

The Soviets also discussed the challenge of comprehension. Moroz and Lev Gindilis of the Shternberg Astronomical Institute lectured on coding and decoding signals. Gindilis took it for granted that some extraterrestrials wanted a dialogue, and he envisioned a game in which both parties desired contact, each clueless of the other and yet bound by the laws of science. B. I. Panovkin of the Soviet Academy of Sciences added that deciphering the symbols of an alien civilization required "a close identity" between the sending and receiving parties. This "stringent limitation" reduced the number of possible correspondents.

For Barney Oliver, comprehension was a semantics problem, secondary to transmission and reception. He described his Project Cyclops, a proposal to erect and connect a thousand or more steerable radio telescopes to get the reception of one huge antenna. The expense was huge, Oliver confessed, but so were the potential benefits. Oliver hoped that by the year 2001 humanity would no longer be isolated but "beginning a new epoch on the evolution of life on earth."

Townes continued pitching the attempt to contact nearby extraterrestrials. Echoing Minsky's earlier comments about technological obsolescence, Townes asked whether today's instruments would be available many millennia in the future, when a response in today's radio code finally arrives from distant Type II and Type III supercivilizations. Townes was making the case for ignoring them. They were too far away.

What were the consequences of contact? Since Marxist ideology had already outlined the future, the Russians prudently allowed the Americans to dominate the conversation. Morrison predicted the enrichment of human thought, comparing alien missives to the discovery of classical Greek learning during the Renaissance (roughly from the fourteenth to the seventeenth century). Morrison used this analogy because Renaissance scholars could not converse with the long-dead Greek literati, a burden similar to astronomers' inability to converse with distant extraterrestrials. Morrison conceded that the alien message would be strange, more likely culture than science, and predicted a long, cooperative effort to make sense of it.

William McNeill had been silent during the technical discussions, but now he spoke up. He was dubious of "the whole notion of extraterrestrial communication," because "deciphering the message" would be very difficult. Between humans and aliens, McNeill discerned few commonalities.

> Our intelligence, unless I misunderstand, is very much a prisoner
> of word, a prisoner of language, and I don't see how we can assume that the language of another intelligent community would
> have very many points of contact with our own. The differences in
> biochemistry, in the sensory span of intelligent beings, sensibility,

in such things as size of body and neuron pattern, all seem to me to add up to a high improbability of mutual intelligibility.

McNeill had yet another stone to throw: "I must say that in listening to the discussion these last days, I feel I detect what might be called a pseudo or scientific religion." Yet McNeill insisted that he was not being negative. "Faith and hope and trust have been very important factors in human life and it is not wrong to cling to these and pursue such faith. But I remain, I fear, an agnostic, not only in traditional religion but also in this new one."

Crick ignored the religious reference and instead challenged McNeill's prediction of mutual incompatibility. Crick pointed to mathematics as "a natural language . . . common to both parties." In reply, McNeill accused mathematicians and natural scientists of suffering from math "chauvinism." "I can't prove it," McNeill continued, "but I don't think you are justified in assuming automatically that our mathematics is commensurate with their mathematics." Oliver pulled rank, dismissing McNeill's remarks as the opinion of the "educated person . . . not intimately acquainted with science." He noted that "another intelligent species very likely also has eyesight" and that pictures could begin a dialogue.

What did the conference achieve? If its purpose was to hasten the day of contact, it failed. It did not identify a contact golden bullet. Whenever a strategy was proposed, for example, searching for wideband instead of narrowband signals, or vice versa, someone found a flaw. The consensus, if any, was that contact was uncertain unless extraterrestrials were broadcasting beacons. Even worse, Minsky pointed out the generational incompatibility of radio equipment. Perhaps most telling was the optimism in the supercivilization discussion. To assume that supercivilizations had not outgrown radio is a peculiar form of presentism. It assumes that very advanced beings retrieve ancient radio technology from their museums to satisfy their curiosity about backward Earthlings.

As for the cultural impact, it could almost be said that Sagan hijacked the conference. The Soviet scientists were the hosts. They wanted to talk about the technology; they got that and a lot more, such as rants on mankind's sorry state. Did the Soviets secretly share this pessimism? Shklovskii probably did. As for the others—no one knows. Did the pessimism of the Western scientists reinforce the Communist view that capitalism was doomed? This much is certain: only the most dedicated Communist could avoid noticing the free-ranging discussion among the Americans.

Today, the transcript of the Byurakan II conference seems eerie. The Soviet scientists were outwardly confident. Who would have guessed that within twenty years their world, the Soviet Union, would collapse? Byurakan II gave the impression that the United States was the doomed superpower. Its best and brightest seemed beaten, so fearful of the future. Appearances truly are deceiving.

In March 1974 the Soviet Academy of Sciences followed up the Byurakan recommendations when it announced an ambitious SETI program that extended to 1990. Apart from ground-based systems to monitor the entire sky, it projected two SETI space stations.[29] These were never built.

The 1981 SETI Conference in Soviet Estonia

In 1981, from December 8–11, the Soviet Union hosted another international SETI conference. Held in Tallinn, Soviet Estonia, it was a pale shadow of the 1971 Byurakan meeting. Although thirty Americans were invited, travel funds were lacking and only ten attended, among them Drake, Oliver, and Morrison; Sagan was absent.[30]

As with the Byurakan II meeting, Drake provided two accounts: a softer one in 1982 and a sharper version in his 1992 SETI memoir.[31] In both narratives, he noted the Soviet media's intense coverage, far more than scientists receive in the United States. Drake marveled, perhaps with envy, at the attention science received in the Soviet Union. Both accounts also noted the depressing winter darkness, the sun unseen till 10 a.m. and gone by 4 p.m. In the second account, Drake added two vignettes. The first involved the female "interpreter" who knew little English yet tried very hard to befriend the American astronomers, and, when ignored, sulked, allowing the real interpreter to take over.

The second featured the food shortage at the two "fancy" hotels, an indication, according to Drake, that the Soviet Union was troubled. The food shortage may have had a more immediate meaning. In the highly centralized Soviet economy, the Soviet establishment allocated food according to priorities; official interest in SETI may have been waning.

In both of his accounts, Drake mentioned the uneven quality of the papers that were presented, 80 percent given by Soviet scientists. Astronomer Woodruff Sullivan (b. 1944) of the University of Washington was likewise critical, describing the papers as ranging from "stale to mediocre to even outrageous."[32] A large number consisted of speculations worthy of American ufology, such as alleged alien visits in prehistory, Martian monuments, and extraterrestrial adjustment of Earth's rotation. Drake claimed that many of the reputable Soviet scientists were embarrassed.

The question naturally arises: how could first-class astronomers such as Shklovskii and Kardashev be associated with ufology? Drake blamed poor peer review in Soviet science, which allowed the "unsound or spurious" to survive. This explanation does not go far enough. At the Tallinn conference, why did the Kardishevs and Shklovskiis permit Soviet ufologists to participate? The answer might be that the Soviet best and brightest did not control the conference; unsophisticated bureaucrats and politicians were the real masters. The greater prestige of Soviet scientists compared to their Western counterparts was illusory; in the area that really counted, the integrity of their disciplines, they had less input.

It may also be that what seemed like ufology was something else. Russian cosmism, never quite dead, was reemerging in the last years of the Soviet Union. In today's Russia, discussion of Fedorov is no longer forbidden, as the N. F. Fedorov Museum-Library in Moscow attests. The Russian Academy of Sciences has been examining cosmism. It sponsors the Institute for Scientific Research in Cosmic Anthropoecology, which has been studying telepathic communication between humans and inanimate objects and also between humans and cosmic forces.[33]

Cosmism aside, the Soviet Union, although troubled, still had bite in 1981. At the Tallinn conference, politics, Soviet-style, intruded. The final resolution contained an innocent-seeming statement "condemning extremist or fanatical position, either pro-SETI or anti-SETI." Although the Americans saw nothing wrong, the Soviets understood these words as code for ideological deviation. Shklovskii was being attacked. Although many Soviet scientists sympathized with him, they all voted for the resolution. Perhaps Shklovskii's fault was his waning SETI ardor. He had delivered a paper proclaiming evolution a near dead end and intelligence a possible rarity in the universe. In claiming that progress had ended, he was denying Marxist optimism. In a later interview, Shklovskii noted that advanced extraterrestrials were not leaving traces—an indication that they did not exist.[34]

By contrast, Kardashev remained on course during the conference, continuing to develop his thesis of supercivilizations. His strategy for finding them was looking for the unusual, especially instances when physical laws keep on failing. Cosmic anomalies could indicate that supercivilizations are modifying their surroundings, Kardashev believed.

Other papers discussed whether radio will best achieve contact. In addition, the Soviets divulged plans for new equipment. Troitskii described a SETI-dedicated array of one hundred one-meter dishes, whose first portion would operate the following year. Kardashev described a new seventy-meter parabolic telescope to be operational within five years.

After the conference, the American astronomers met with Kardashev and Troitskii to discuss cooperation. Although the Americans had the computer skills, the Soviets had the telescopes. As Drake later noted, American SETI was in a sad state, with Congress hostile to public funding.

Yet the future of SETI belonged to the United States. American SETI eventually left NASA, went private, and in the new millennium it survives. On the other hand, the Soviets never built the dish array. Like the Soviet Union itself, Soviet SETI declined. SETI has been little heard from in the Russian Republic that has replaced the Soviet Union. It presumably bores Vladimir Putin.

The Enigma of Soviet SETI

Soviet SETI was theoretically impressive but culturally sterile. When Soviet SETI scientists met, they kept to a strict no-nonsense discourse that rarely strayed from technical questions. Unlike their American counterparts, who often revealed their very human motives, Soviet SETI people rarely expressed their true feelings. Whereas Sagan sought reassurance of humanity's survival and Morrison wished for cultural enrichment, the hopes and desires of the Soviet SETI scientists are unknown. Did they seek contact only to boost the prestige of the Soviet Union and Mother Russia? Can they be dismissed as foot soldiers in the Cold War?

Shklovskii had a wider vision; he wanted reassurance that extraterrestrials had survived. When the few SETI searches suggested that the galaxy was not full of intelligent life, he lost heart and concluded that extraterrestrials had destroyed themselves, just as humans would soon do. What did the other Soviet scientists think? Did they agree with Shklovskii, or did they have an irrational belief in something great out there, fueled by the Russian cosmism tradition. Were they later versions of Tsiolkovsky, romantic philosophers as well as scientists? Whatever their private thoughts, in public they were silent. Unlike American SETI, Soviet SETI hid its personal context when facing the world. Like the Soviet Union itself, Soviet SETI was dull.

7

The 1970s

Gaining Respect

When the Soviet Union launched the first Sputnik satellite, in 1957, the space race started. In the following years, when the Soviet Union launched first-class satellites and rockets, many nervous Americans feared a military strike from space. By contrast, the Soviet lead in SETI did not frighten, for there was no SETI race. Americans did not besiege Washington with letters and phone calls urgently demanding that the United States be the first to contact extraterrestrials. NASA had no pressing reason to develop a SETI program.

John Billingham

If SETI were to be planted and thrive in NASA, it needed a savvy advocate. The SETI community had such a person. He was John Billingham (1930–2013), who ran NASA's space medicine unit. A NASA insider, Billingham stood out from the other SETI pioneers, who were mostly academicians, more comfortable in the classroom and the laboratory than in the labyrinths of the Washington bureaucracy.

Billingham's biography was not typical SETI. Unlike Drake and Sagan, science fiction had not inspired him. In his youth he had ignored the genre, never having read Olaf Stapledon, the mystical bard of future space travel. Besides, by SETI standards, Billingham had an unusual occupation. Unlike the early SETI notables, who were engineers like Oliver and Bracewell or physical scientists like Morrison, Drake, and Sagan, Billingham was a medical doctor as well as a bureaucrat. A graduate of Oxford University, Billingham was not even American by birth.

Born in Worchester, England, Billingham received a medical degree from Oxford in 1954 before being drafted by the Royal Air Force and spending eight years at the Royal Air Force's Institute of Aviation Medicine. Here he studied the physiological stresses of extreme flying conditions. Meanwhile, the Russians had launched Sputnik, prompting the British Interplanetary Society to hold a conference on human space flight. Billingham presented two papers: one on surviving the harsh conditions of the moon, and the other on life support in spacecraft. Only a leap of the imagination could equate the rigors of space with aviation medicine. Billingham was straying from his day job. His superiors were not all that pleased.

Billingham soon became known in American circles. The Royal Air Force was working closely with its American counterpart, and Billingham represented the RAF at joint scientific meetings. At the same time, the American space program was going into high gear. Space had replaced jet travel as the research vanguard. Billingham could not resist being part of that vanguard. In 1963, Billingham contributed to the "brain drain" from Great Britain when he accepted a NASA offer. Given the choice of the Johnson Space Center in Houston or the Ames Research Center in Mountain View, California, Billingham accepted Houston, for in his words it was "clearly going to the moon." He was chief of the Environmental Physiology Branch. Nonetheless, in 1965, he switched to the Ames Research Center, having received an offer he "could not refuse," doing the same thing as at Houston, except that Ames was "research oriented, looking into the future."

Billingham recalled that the top floor of the life sciences building at Ames housed "strange and interesting people," about forty or fifty exobiologists who were devising experiments to detect non-terrestrial life. In conversations, Billingham learned of Sagan and Shklovskii's SETI classic. After reading it cover to cover, he "sat back and said 'Wow!'" Transformed into a SETI true believer, Billingham realized the limited vision of the Ames exobiologists. They were seeking alien microbes when an intelligent universe might be out there.[1]

American SETI had a long way to go. After Drake's Project Ozma, American SETI had languished during the 1960s, limited to a few research papers and discussions. In 1971, Gerrit Verschuur (b. 1937) used the Green Bank Telescope to examine ten nearby sunlike stars over a two-year period. Verschuur was the second American to do a SETI search. Compared to Drake, Verschuur compiled far more data, which he duly published.[2]

Like Drake in Project Ozma, Verschuur had limited time on the Green Bank Telescope. Only NASA could assure SETI of dedicated resources. But first a thorough search plan, one that specified the necessary equipment, was necessary. Billingham decided to do something— to replace thought and talk with action.

Project Cyclops

Luckily for SETI's future, Billingham had been hosting a program in engineering systems design. Every summer, in conjunction with Stanford University, about twenty university faculty spent three months at the Ames Research Center working on a design problem, something exotic but still close to present reality, such as designing a moon base. When Ames got a new director, Hans Mark (b. 1929), in the summer of 1969, Billingham proposed making SETI the topic for the 1970 summer program. Although intrigued, Mark saw the pitfalls surrounding SETI. He counseled a gradual approach, doing a small study in 1970 and, only if feasible, the definitive study in 1971. Billingham accepted his boss's advice.

The next step was choosing a leader for the 1971 study. Two obvious candidates were Frank Drake and Barney Oliver. Drake was the only American to have done hands-on work on SETI, but Oliver was chosen. Drake was an astronomer, whereas the project needed an engineer to design technology for detecting extraterrestrial signals. Oliver was an electrical engineer as well as a SETI partisan, and he had another virtue: in 1966 he had proposed a detection system that prefigured his 1971 study, his classic Project Cyclops.

In the summer of 1971, Oliver took a three-month leave from Hewlett-Packard to work on Project Cyclops. He had at his disposal the twenty-four scientists and engineers in the summer study group together with consultants from academia, industry, and government. The result was a 243-page report, "The Design of a System for the Detection of Extraterrestrial Intelligent Life" (NASA's Contractor Report no. 114445). Billingham, the coauthor, claimed Oliver did most of the work. For his part, Oliver was modest, calling the finished product "very preliminary" and needing additional work before the "final detailed design." It was good enough. In the next ten years more than fifteen thousand copies were distributed, a best-seller by engineering standards. Billingham later called it "the foundation of everything that has happened since."[3]

What exactly was Project Cyclops? According to the report itself, its primary intention was identifying the necessary "hardware, manpower, time and funding to mount a realistic effort" to detect extraterrestrial intelligent life. That hardware included the world's largest radio telescope.[4] Project Cyclops was the first fully integrated examination of SETI. In addition to presenting the specifications for the proper search equipment, it made the case for a search in the first place, a case that Drake, Morrison, Sagan, and others had already made, but now summarized and combined with a realistic technological blueprint.

Oliver was a radio-wave chauvinist. He rejected probes as too costly and space flight as too futuristic. He dismissed Kardashev for equating "advancement with mastery over energy resources." Oliver saw no point in looking for Dyson spheres. The principle of Occam's razor, which sees simpler explanations as preferable, suggested to Oliver that extraterrestrials had contained population through a simpler method, such as birth control.

Oliver insisted that these contact strategies suffered from still another flaw; they implied human passivity. Humans should not wait until Bracewell's probes arrived or until new technologies detected Kardashev's supercivilizations or Dyson spheres. Oliver called this waiting strategy a laziness unworthy of the human spirit.

By contrast, Oliver noted, Project Cyclops was proactive; it did not gamble that extraterrestrials had embarked on one particular path of development. Nonetheless, Oliver was also gambling. In Oliver's vision of the universe, aliens might or might not have sent probes, built Dyson spheres, or created supercivilizations, but they were transmitting radio messages. Oliver was betting that extraterrestrials, unlike Earthlings, had not mastered a communications technology using other portions of the electromagnetic spectrum. If they had mastered this technology, Oliver was betting that his advanced extraterrestrials were dumbing down, using Earth's twentieth-century radio technology so that we fortunate Earthlings could receive their message.[5]

If Earth was unlucky and extraterrestrial civilizations were clueless of other worlds but curious, all was not lost. Extraterrestrials could be randomly transmitting a beacon in all directions, hoping for a lucky hit. Since a beacon was a deliberate attempt to communicate, Project Cyclops took it for granted that transmitting races were making "the job of deciphering and understanding the messages as simple and foolproof as possible." Oliver foresaw a pictorial message, since an intelligent race would "very likely have vision."[6]

Oliver admitted that extraterrestrials might not be transmitting beacons, forcing SETI to eavesdrop on the chatter between extraterrestrial civilizations as well as the internal chatter of individual civilizations: "leakage" in astronomers' lingo. Leakage was the extraterrestrial version of radar or television transmissions (assuming extraterrestrials were not using cable). Leakage transmissions would be weaker than beacons and would require powerful receivers. Yet leakage could not be ignored. A thorough search of the heavens required a system to detect weak leakage signals as well as powerful beacons.

According to its designers, success required an antenna three to five kilometers wide, making it the world's largest telescope, dwarfing the three-hundred-meter dish at the Arecibo Observatory in Puerto Rico. The designers knew that a telescope of this size was out of the question. Instead of a massive dish, the final blueprint envisioned a thousand small antennas, perhaps as many as twenty-five hundred, each about a hundred meters across. The array would cover an area ten kilometers in diameter, equivalent to sixty-five square kilometers.

Cyclops illustrated the old saw about the power of numbers. Although the individual antennae were not huge, their combined and processed data would detect beacons from as far as one thousand light-years in space, an area that included a million sunlike stars. The huge array would also catch weaker signals, the leakage from the extraterrestrial version of television and radar.

Unfortunately, Oliver and his staff, like all SETI people, were ignorant of ET's radio frequencies. After studying the electromagnetic spectrum, they concluded that the portion worth close examination lay in the frequency range of 1,000 to 3,000 MHz. Here lay more than enough work: about two billion channels, which had to be examined from different directions. Clearly, a shortcut was needed; Oliver had to narrow the range. In their seminal paper, Morrison and Cocconi had proposed a specific radio frequency, 1,420 MHz, a "quiet" part of the radio spectrum. Oliver instead proposed a band of frequencies. According to the Project Cyclops report,

> Nature has provided us with a rather narrow quiet band in this best part of the spectrum that seems especially marked for interstellar contact. It lies between the spectral lines of hydrogen (1420 MHz) and the hydroxyl radical (1662 MHz). Standing like the Om and Um on either side of a gate, these two emissions of the disassociation products of water beckon all water-based life to search for its kind at the age-old meeting place of all species: the water hole.

This passage reveals another side to Oliver. Scientist and man of business, he was also a romantic. Seeing the span between the spectral lines of hydrogen and the hydroxyl radical as a water hole was a touch of poetry. Over the years, astronomers have proposed other "magic" frequencies, each touted as the one extraterrestrials were likely using, but in truth, alien minds are impossible to fathom.

Although intended as a receiving station, the Cyclops array could also transmit to other civilizations. Since sending and receiving could not occur at the same time, Oliver suggested first using the array as a receiver and if the nearest thousand stars did not signal, the array could try beaming, in case ET had radio but was not transmitting. If no response came, the array could return to its receiving mode.

Apart from the array, Project Cyclops also entailed staff facilities.

Instead of commuting from afar, workers would live nearby in a special town called Cyclopolis. Oliver admitted, though, that staff housing had not been finalized. Instead of Cyclopolis and its short commute, he was open to having "the necessary houses, stores, schools" at the central headquarters. Furthermore, "playgrounds and recreation facilities" could fill the space between the antennae. Oliver was envisioning a science city whose attention to detail had a touch of Soviet Union planning. The very scale as well as the art deco touch to the name "Cyclopolis" has led to the accusation that Project Cyclops had "a quaint hubristic madness about it."[7]

The real problem with Project Cyclops was not ambition and pride, but money. Its projected cost over the next ten to fifteen years ranged from $6 billion to $10 billion (in 2014 dollars, $37 billion to $61 billion).[8] To soften the impact of this staggering cost, Oliver noted that not all the money had to be spent up front. If only one hundred antennae were built each year, the annual cost would sink to about $600 million (roughly $3.7 billion in 2014). Oliver emphasized that the array could operate before completion. If they caught a break and achieved contact quickly, expansion of the array could stop. Oliver believed in that lucky break. At the beginning of the report, he quoted from Frank Drake's 1962 SETI book, in which Drake stated his "almost absolute certainty" that intelligent beings were transmitting radio waves to Earth.

Although a true believer, Oliver was honest. He admitted that the search could take "decades and possibly centuries." Project Cyclops needed "a long term funding commitment" and "faith" that the quest was worth the cost, "faith that other races . . . have been equally curious and determined to expand their horizons." Besides, argued Oliver, "the quest for other intelligent life" had popular appeal; it might receive "support from those critics who now question the value of landings on 'dead' planets and moons."

Ultimately, the case for Project Cyclops rested on the case for SETI. Oliver defined SETI as "a legitimate scientific undertaking," deserving inclusion in "a comprehensive and balanced space program." In support, he gave the usual SETI arguments. He explained the mediocrity principle, pointed out the knowledge and wisdom gained from the Galactic Club, and linked human survival to contact.

As for culture shock—the human reaction upon learning other creatures were smarter—Oliver conceded that it was possible. He was unfazed, noting that contact would not be face-to-face but through radio, allowing the vast distances of space to cause "long delays" and slow exchanges of information. Humanity had time "to adapt to the new situation." "After all," Oliver wrote, "generations might be required for a round trip exchange" of radio messages.

NASA Rejects Project Cyclops

Researching and writing the report was easy. Selling it to NASA's higher echelon was something else. SETI had the bad luck of receiving a weak recommendation from the National Science Foundation's decennial blueprint for American astronomy. Written by Jesse Greenstein (1909–2002), chair of the Caltech Astronomy Department, and issued in 1972, the *Report of the Astronomy Survey Committee* approved of SETI, giving it some legitimacy. This was the good news. On the other hand, the five-hundred-page Greenstein Report devoted only a few pages to SETI. Although calling for "new funds for what may become a new science," the report proposed no spending figures as it did with other projects.[9] Obviously, SETI was defined as a very modest pursuit. The big bucks of the expensive Project Cyclops were anything but modest.

The NASA administrator was James C. Fletcher (1919–91). He would be on the firing line in case Congress or the public adversely reacted. The official cheerleader for the space program, Fletcher believed that most American scientists preferred spending on other areas of science, that is, their specialties. He also believed that the general public opposed the nonmilitary part of the space program.[10] Persuading a person with these views to adopt Project Cyclops would be tough.

In internal correspondence, Fletcher liked Project Cyclops in principle but found the price tag daunting, "particularly in view of the current climate," with the federal government downsizing its space program. He was convinced that Project Cyclops could not stand alone. It needed to be part of "an overall project in astronomy and cosmology."[11]

Project Cyclops's prospects remained bleak on September 11, 1973, when Fletcher and other NASA functionaries met with Oliver, Billingham, and Hans Mark. Fletcher deemed it "undesirable" to fund a program that after ten years might not have produced "significant results." Fletcher prodded Mark and Oliver to offer positive assurances of contact; they could not, although they did offer a high probability of contacting a million-year-old civilization. Fletcher rejected this imprecision, stating that the public (a euphemism for Congress?) would reject the "long shot" of an alien civilization having this longevity.[12] Fletcher was reflecting the NASA mind-set, which Billingham later defined as creating "beautifully worked out and orchestrated" projects, planned to the second and exuding certainty. By contrast, SETI embodied uncertainty with contact taking place in the near or distant future, if at all.

A few days later, Oliver wrote to Fletcher and again pitched for Project Cyclops. He insisted that the array had a 100 percent chance for success. Fletcher was dubious, rejecting Oliver's belief in several extraterrestrial civilizations surviving for a million years. On Earth, Fletcher observed, the long-lived Byzantine empire had lasted only a thousand years, although it had "many stable institutions."[13]

Oliver tried again, this time sending Fletcher a fourteen-page memorandum. Oliver did not hold back. He cited Blaise Pascal, the French mathematician and philosopher of the seventeenth century, who had famously said that the possibility of eternal life justified Christian faith, although heaven was an uncertainty. Oliver claimed that the probability of contact was greater than that of eternal life. He also threw in the Cold War, contrasting the low cost of Cyclops with the "wasted" effort of defense spending. Besides, added Oliver, recent and projected advances in receiver and computer technology were reducing the cost of Project Cyclops.[14]

Barney Oliver never stood a chance. He was dealing with a bean counter, Fletcher the bureaucrat, who had to account for the public's money. Failure is always difficult to justify. Justifying an expensive extraterrestrial search with no guarantee of success, and, even worse, with much of the public confusing it with ufology, was well-nigh impossible.

For his part, Oliver in his day job was as hardheaded as they come. Only a person with a cold view of the possible could direct research at Hewlett-Packard, lead the team that developed the pocket calculator, and hold fifty patents. Oliver liked to parade his realistic bent. An avowed atheist, he saw no tangible evidence for God and dismissed the afterlife as "mythology." Yet his belief in extraterrestrials was akin to a walk on the wild side, as irrational as the religion that he rejected as beyond logic.[15]

Oliver had fought the good fight, but Hans Mark had already conceded defeat when in lieu of Cyclops he proposed a two-year $500,000 appropriation so that NASA could further study the probability of contact as well as the non-SETI applications of the Cyclops array.[16] Project Cyclops was not dead, but as of 1974 it was slowly dying.

The verdict on Project Cyclops, Morrison later said, was that it made SETI searches appear expensive.[17] Yet, as it turned out, a huge array was not necessary. Oliver did not foresee Moore's Law: computing power doubles every six months. If he had known this, his proposal would have projected lower costs and stood a better chance. Whatever the merits of Project Cyclops, it cannot be denied that for the first time an official government document contained the potential benefits of contact.

The Boston SETI Symposium

Fear of the public's skepticism helped to sink Project Cyclops. SETI partisans needed to educate and win over the public as well as the scientific community. On November 20, 1972, NASA co-sponsored a symposium on "Life beyond Earth and the Human Mind" at Boston University. Richard Berendzen (b. 1938), chairman of the Boston University Astronomy Department, moderated. Well attended and reported in the *New York Times*, the symposium featured Sagan, Morrison, Harvard professor and Nobel laureate in medicine George Wald (1906–97), theologian Krister Stendahl (1921–2008), and anthropologist Ashley Montagu (1905–99), famous for his writings on race and gender. If the organizers chose this panel for intellectual diversity and outreach to a lay audience, they succeeded. Unlike the Byurakan conferences, the panelists spoke in language that educated non-scientists would understand.[18]

Nikolai Kardashev had wanted to exclude "windbags" from the Byurakan conference. If he had attended the Boston symposium, his reaction to Ashley Montagu can only be imagined. Montagu had not prepared for the conference, knew little of SETI, and concealed his ignorance with moral superiority worthy of a Savonarola. He obviously had a chip on the proverbial shoulder, having lost his academic position at Rutgers University during the Red Scare of the 1950s. Montagu was hypercritical not only of politicians but of democracy itself. Americans will be shocked, he said, to find out that extraterrestrials do not elect their politicians but require "all candidates to be not only knowledgeable but also loving beings, who are appointed to office only after they have passed the most rigorous of examinations." Humanity was so debased, Montagu said, that extraterrestrials no doubt dismissed it as "rabies or cancer or cholera—in short, as a highly infectious disease best quarantined from the rest of the universe." Montagu worried that humans might contaminate extraterrestrials. He foresaw the Nixon administration teaching extraterrestrials the art of making peace through war. Humanity was sick and dangerous, Montagu concluded; it worshipped "false values" and was not ready to meet extraterrestrials.

No hellfire preacher could have delivered a better sermon. However, Montagu was not in church but at a SETI conference. By insisting that contact would ruin extraterrestrials, he had unwittingly made a great case against SETI. He was obviously ignorant of the contact process and oblivious of the meaning of the term *extraterrestrials*, using the clumsy circumlocution "beyond earthers." Montagu did serve as a useful straw man; he gave Morrison a pretext to describe the SETI basics: the vast distances of space prevent physical contact and make radio conversations very lengthy, perhaps spanning generations.

George Wald was also uneasy. Unlike Montagu, Wald did not worry whether contact would harm extraterrestrials; he feared the effect on humans, the shock to human self-esteem once advanced knowledge was passively received. "One could fold the whole human enterprise," Wald said, "the arts, literature, science, the dignity, the worth, the meaning of man—and we would just be attached as by an umbilical cord to that 'thing out there.'" Berendzen replied that cancer patients will not care whether their cure comes "from the Boston Medical Center or from Tau Ceti." Morrison took the long view: deciphering and understanding an alien message will take decades, he said. Predicting a gradual effect on humanity, he was "neither fearful nor terribly expectant."

When Berendzen mentioned a cancer cure, he was not breaking new ground. He was echoing Drake's words in the days before Project Ozma. In both cases, there was a belief in an essential equivalence, the stuff of science fiction in which humans and aliens mate. The best question to have asked at Boston was not whether humans should accept an alien cancer cure, or even whether aliens will graciously provide it, but rather if aliens suffered from cancer at all. Selling SETI, however, required a belief in that cure.

Post-Contact Bliss

By coincidence, the year of the Boston symposium saw the publication of *The Listeners*, a science-fiction novel by James Gunn (b. 1923). Like the symposium, the novel pushed the message that contact will hugely enhance humanity. The agent of this sea change in Gunn's novel is an extraterrestrial message. Humanity replies, and while waiting ninety years for the alien response, the human race becomes increasingly aware of its oneness and learns to live in peace—in effect, it achieves world government, the dream of the atomic scientists of the 1940s and 1950s. Gunn derived his utopian theme from the writings of SETI thinkers such as Morrison, Sagan, and Drake, whom he liberally quotes.

In a 1976 article, Drake gave his version of post-contact bliss, surpassing Gunn's *The Listeners*, by claiming that contact will lead to human immortality. Drake speculated that some extraterrestrial races are immortal, having discovered how to stop aging, repair its ravages, or transfer memories to a clone. These extraterrestrials die only if their bodies are destroyed. Hence, Drake insisted, they are obsessed with avoiding accidents and violence. First, they have eliminated war among themselves, and second, they seek to "avoid threats from another planet" by spreading the secrets of immortality so that potential invaders will also be obsessed with personal safety and renounce war.

In short, according to Drake's reductive reasoning, the immortals will teach humans how to live forever. If gods are defined as immortal, one can infer from Drake's analysis that men will be gods, or at least a facsimile, after contact is made.[19]

Are Drake's immortals happy? In *Gulliver's Travels*, Jonathan Swift wrote of the "struldbrugs," immortals who are bored, lacking urgency since they can defer every decision to the future. By contrast, Drake's immortals have a purpose: preventing body-destroying accidents. As a result, they construct "every device and vehicle . . . as to present no lethal hazard under any circumstance. . . . The use of aircraft for transportation, or indeed for any purpose, might be impossible, since a falling aircraft is a hazard to those on the ground as well as those aboard." Did Drake realize the price paid for refusing to risk accidents? His immortals are dull. They have banished from their lives the greatest source of excitement: the fear of death. Why else do people skydive and climb Mount Everest? Drake's immortals have no epics of war and passion; no *Iliad* is composed, no *War and Peace* is read. Drake's immortals are technological hypochondriacs, almost comical, and his article may have backfired. Philip Morrison later called it "very tendentious" and "a small mistake."[20]

Pioneer 10 and Pioneer 11

Drake also promoted SETI in his collaborations with Carl Sagan. Thrice in the 1970s, Sagan and Drake conflated their astronomical work with sermons to the benighted, but they preached gently and without Montagu's misanthropy. The first time was when they attached plaques to the Pioneer space probes of 1972 and 1973, then again when the Arecibo radio telescope beamed a message to the Hercules Constellation, and finally in the Voyager probes of 1977, for which they and their team selected pictures and recorded sounds of Earth. The alleged purpose of these well-publicized endeavors was explaining Earth to extraterrestrials. As Drake candidly admitted, the real purpose was sending a "message" to humanity.[21] The public was reminded of the space program, of extraterrestrials, and as a bonus given a mild dose of 1970s progressivism.

NASA launched Pioneer 10 on March 2, 1972, and Pioneer 11 on April 6, 1973. Pioneer 10 was programmed to fly by Jupiter, send back data and photographs, and eventually leave the solar system. Pioneer 11 had roughly the same mission, except that after Jupiter it flew by Saturn before its endless journey into outer space. Pioneers 10 and 11 were the first human objects destined to leave the solar system. Although the probes would take eighty thousand years to reach the nearest star, there was the faint possibility of extraterrestrials' finding them. When a *Christian Science Monitor* writer suggested that a message be included on the Pioneers, Sagan was enthusiastic, hoping beyond hope to win the SETI lottery. NASA gave Sagan three weeks to devise drawings for a six-by-nine-inch gold anodized aluminum plaque.[22]

Sagan and Drake's message was partly innocuous and partly controversial. The safe part was Drake's "map." It positioned the sun relative to the center of the galaxy and fourteen pulsars (neutron stars), shown as lines with binary numbers. Since the fourteen pulsars are observed only from Earth, it was hoped that extraterrestrials will understand this and pinpoint Earth's location. The fourteen pulsars served a second purpose. Pulsars emit radio waves at known rates, each of which has its own characteristic frequency. This makes pulsars a time clock of sorts that will reveal when the Pioneers were launched. Another part of the map described our solar system; the map placed the nine planets in relative distance from the sun.[23]

Sagan insisted that aliens with an advanced technology could read Drake's map, since it was written in the "only language . . . share[d] with the recipients: Science." Although he was quite willing to concede that aliens and humans totally differ in appearance and body chemistry, Sagan nonetheless held that human and alien minds process data much the same way, allowing very physically different beings to decipher the scientific notation of humans. Sagan never perceived the contradiction in this reasoning. Yet, as with the whole SETI enterprise, what else could he and Drake do? They had to hope for the best, that is, convergence of thought between extraterrestrials and humans.[24]

Few, if any, on Main Street cared whether ET could decipher the Pioneers' scientific notation. The plaques' depiction of a man and a woman was something else. The two were naked, in full-frontal nudity, nothing hiding the naughty bits. The whole thing shocked many Americans, leading to the question whether nudity was really necessary. Clothes would have certainly puzzled extraterrestrials if they lacked prudery or their benign environment made skin covering unnecessary. On the other hand, if they wore clothes, human nakedness might have been unsettling. A newspaper cartoon of the day features a human-looking couple on Jupiter finding the plaque. Dressed in a business suit and tie, the man tells the woman that the Earthlings are "like us" but they don't wear clothes.

Linda Salzman, Sagan's second wife, drew the figures. Carl Sagan was in his preaching mode. He once blamed society's ills on sexual repression, asking whether "severe sexual repression" was a crime against humanity.[25] The nude figures on the plaques were Sagan's way of telling the world that nakedness and sex were natural. It is telling that the plaque did not depict children. Sagan was not suggesting reproduction, only the behavior that causes it. The angry letters to newspapers accusing NASA of sending smut into space may have been written by bluenoses, but these prudes nonetheless sensed the drift; they got Sagan's real message.[26]

Sagan had feared NASA's reaction when he proposed the nude drawing. He later added that he should not have worried. He was surprised, though, when identity groups were offended. Feminists complained of female subservience: the man on the plaque was waving his right hand in a friendly greeting, whereas the woman did not wave. Gays noted the couple in the drawing was heterosexual and accused NASA of homophobia. Still others objected to the race of the two figures.[27]

This much is clear. Sagan had tried to send a message to Earth. In reality, Earth had sent a message to Sagan, and he missed it. Earth was telling Sagan that no one can fully comprehend a totality. Sagan foresaw traditionalists being offended. He did not foresee identity groups taking umbrage. The same incomprehension may hold for Sagan's belief in extraterrestrial communication. No human may be able to penetrate the uniqueness of the extraterrestrial mind (or what passes for it). At best, humans may understand only bits and pieces of an alien message, as if it were a text in a lost alphabet of a dead language. A few symbols are deciphered, but the full meaning remains forever elusive.[28]

Sagan and Drake at Arecibo

In 1974, Sagan and Drake did it again, when the Arecibo Observatory was upgraded. A new reflector surface and a new transmitter capable of sending a signal thousands of light-years into space deserved a dedication. Sagan and Drake used the event to transmit an interstellar greeting in binary. Using 1's for dark spaces and 0's for blank spaces, Drake depicted chemical formulas, diagrams of the molecules essential for earthly life, and figures of a human, the solar system, and a radio telescope roughly like Arecibo's.[29] Human-centric, Drake's message presumed that extraterrestrials could both see and decipher the images. On November 16 the Arecibo telescope transmitted the greeting to the Hercules constellation, known as Messier 13, about 25,000 light-years into space.

What could go wrong? This time Drake and Sagan had presumably avoided controversy. Not so! Upon hearing the news, Sir Martin Ryle, Nobel laureate and Great Britain's Astronomer Royal, questioned the wisdom of revealing Earth to extraterrestrials who could be powerful and evil. Ryle asked astronomers to stop announcing the human presence.[30] Sagan and Drake replied that no harm had been done, since television, radio, and radar had already announced humanity to the galaxy.[31] Drake and Sagan could have mentioned their prudence; they did not beam to nearby stars. The distant cluster they chose will receive the message 25,000 years in the future. As with the Pioneer probes, the real recipient of the message was Earth itself. In taking extraterrestrial existence for granted, Drake and Sagan were validating SETI.

Having transmitted, why not listen? Drake and Sagan returned to the Arecibo Observatory during 1975 and 1976. They scanned four nearby galaxies, hoping to find Type II or Type III civilizations. The entire search was brief, consuming no more than one hundred hours of telescope time. They detected nothing.

Voyager 1 and Voyager 2

Sagan and Drake went back to the preaching mode with the Voyager 1 and 2 probes, launched on August 20 and September 5, 1977, respectively. The probes had two missions. The first was the more important: exploring the outer solar system from Jupiter to Uranus. The second mission was to leave the solar system for an endless journey into the Milky Way galaxy. As with the Pioneers, there was a slight chance of extraterrestrials' retrieving the Voyagers. In the case of that happy event, both Voyagers had affixed a gold-coated copper phonograph record containing 118 photographs and, according to Sagan and his team of scientists, artists, and musicians, "almost ninety minutes of the world's greatest music; an evolutionary audio essay on 'The Sounds of Earth'; and greetings in almost sixty languages (and one whale language), including salutations from the President of the United States and the Secretary General of the United Nations."[32]

Like the Pioneers' plaques, the pictures and recordings in the Voyagers were a message to Earth—in part conscious, and the rest probably beyond the awareness of the Sagan/Drake team. As could be expected in a greeting to extraterrestrials, the photographs included views of the solar system, the planets, and Earth. Insects and animals represented non-human life. Human reproduction received much attention, beginning with DNA, followed by conception, the growth of embryos, birth, and finally nursing mothers. To complete the cycle of reproduction, it was necessary to show how embryos begin. A photograph of a nude couple holding hands was proposed to NASA. Although Sagan described the photo as "extremely tasteful," NASA, obviously recalling the flap over the Pioneers' nudes, rejected it.

About 40 percent of the 118 pictures showed people of all races, working, eating, playing, and being themselves. Other pictures showed off human civilization. The Taj Mahal, the UN Headquarters in New York City, a New England frame house, and Boston as seen from the Charles River displayed the diversity of building technology. Trains, planes, and roads demonstrated mechanical means of transportation.

The photo and music selection was intentionally incomplete. To avoid a bad impression, photographs of "war, disease, crime, and poverty" were deliberately excluded. Religion was another taboo subject. The alleged reason was avoiding controversy, since all religions could not enjoy "equal time." Art was also missing, allegedly due to insufficient time to select the world's best.

The rest of the record contained a potpourri of sounds, beginning with a greeting from U.N. Secretary-General Kurt Waldheim, followed by greetings in fifty-two languages. Also included were over thirty non-human sounds: whale songs, animal cries (among them hyenas and birds), natural noises such as volcanoes and rain, and the artificial noises of technology such as Morse Code, train whistles, and the hum of a sewing machine. The record ended with human music, and here the choices were arbitrary. Samples from the Western canon of classical music were a given, but which composers? Bach and Mozart were chosen. They were joined by a Pygmy girls' initiation song, a raga from India, pan pipes from Peru, Mexican mariachi, and many more selections. The only American musicians making the cut were Chuck Berry, Louis Armstrong, Blind Willie Johnson, and the anonymous singers of a Navajo chant. So much for Cole Porter and Elvis Presley! The Sagan/Drake team brilliantly foresaw the future; their selections meet today's canons of diversity.[33]

If aural beings find the Voyagers' records, they might try to decipher the sounds. They might assume they are listening to an organic whole in which the sounds of the hyena, the train whistle, and Louis Armstrong's voice are related. There is a good chance that this cacophony of random noises will thoroughly confuse extraterrestrial linguists and cryptographers, who will take for granted they are dealing with connections and a common thread.

It may be, though, that highly advanced beings will react less to the Voyagers' random noises than to the Voyagers' primitive technology. When Voyager 1 left the solar system on August 25, 2013, its technology was more than thirty-six years old. Its copper records are obsolete. Voyager 1 had less computing power than an iPhone of 2013.[34] Extraterrestrials may be unable to decipher the copper records. If they can, they may conclude that Voyagers' creators were technological troglodytes and dismiss the "random noises" on the records.[35]

More than anything else, the Voyagers' sounds and images were a sermon to humanity. They are a normative view of things, the ideal world as envisioned by highly educated American progressives of the 1970s. The omissions are telling. Scenes of war and violence were omitted because they were might be threatening to extraterrestrials. This may be so, but the Sagan/Drake team was also suggesting that war is abnormal; mentioning it would be a validation of sorts. Sagan in his maturity was never shy about his pacifism. Another missing topic was politics, perhaps because they reflect the nation-state, whereas Sagan was a self-conscious citizen of the world, who had transcended petty patriotism.

As for the omission of religion, the diversity of faiths is a poor excuse. Christianity, Islam, Buddhism, and Hinduism encompass most of humanity. Ignoring religious song and music mocks the multicultural pretensions of the Voyager records. The Islamic call to prayer is perhaps the most frequently heard human sound. Church bells, although less heard today, nonetheless define many areas of the world. Besides, every selection in the Voyagers was a choice and potentially insults the omitted. The Golden Gate Bridge and the Sydney Opera House are shown, but the Great Pyramid is absent, possibly offending Egyptians.

If the Sagan/Drake team feared controversy, why did it propose a picture of a naked couple to NASA, knowing well the brouhaha over the Pioneers' nude drawing? Given Sagan's liberal views toward sex, the picture was a not-too-subtle commentary on the remaining vestiges of Victorian prudery. The picture showed that the Sagan/Drake team was unafraid of controversy. Religion was omitted from the Voyagers because Sagan and Drake were freethinkers, who omitted religion to deny its validity.

If defining humanity required displaying the many phases of reproduction, why omit its antonym, death? Why omit defecation? These define living creatures even more than reproduction, since many individuals fail to procreate, but everyone defecates and will surely die. The answer is simple. Our age regards death and defecation as impolite subjects. If pornography is something unmentioned in genteel company, death and defecation have replaced sex for what passes as pornography. It may be that Sagan and Drake were more culture bound than they could see or admit.

The individual contents of the Voyagers' record seem innocent enough, but their totality breaks with tradition to suggest a new paradigm for humanity. They express a mood caught well in the lyrics of John Lennon's modern hymn, *Imagine*.

Imagine there's no heaven, it's easy if you try
No hell below us, above us only sky
Imagine all the people
Living for today

Imagine there's no countries, it isn't hard to do
Nothing to kill or die for and no religion too
Imagine all the people
Living life in peace

You may say I'm a dreamer
But I'm not the only one
I hope someday you'll join us
And the world will live as one

The question naturally arises: did the Voyagers' message have much influence? According to the *Readers' Guide to Periodical Literature*, little was published on the Voyagers' images and sounds. The *New York Times* limited its coverage to several articles from its science reporter, John Noble Wilford, and a feature from Ann Druyan (b. 1949) in the Sunday magazine. Both Wilford and Druyan dealt with specific pictures and musical pieces in the Voyagers—not the omissions: the missing sounds and images of misery, politics, and religion. Only a careful reader might notice the omissions and grasp the subtle, humanistic message. The Voyagers' sound and picture show was a case of elites speaking to each other. Like the plurality discourses of previous centuries, the message of the Voyagers aimed at the educated upper crust and bypassed the masses.

Apart from the trendy humanism, Sagan and Drake were making still another statement, one directed mostly at NASA and at astronomers beyond Washington's beltway. These were reminded that extraterrestrials existed (it is hoped), thereby implying the legitimacy of SETI. These reminders were necessary, because few American SETI searches were being conducted. After Gerrit Verschuur's follow-up to Project Ozma in 1971, there had been only an on-and-off series of searches, in all over five hundred hours, at the Green Bank Observatory.

The "Wow!" Signal

The continuous SETI work at the Ohio State University Radio Observatory was perhaps more important. Although not sophisticated, the Ohio State SETI operation may have received an alien signal in 1977.

The observatory had begun as traditional astronomy—sky mapping—when John Kraus (1910–2004) of Ohio State University built a large radio telescope, dubbed "Big Ear," in Delaware, Ohio. In 1972, Kraus's funding abruptly ended. Kraus's associate, Robert Dixon, proposed using the equipment to search the "water hole" for narrowband signals from extraterrestrials. By late 1973, Big Ear was listening. At first it used an eight-channel receiver, and eventually it moved up to fifty channels. Compared to Oliver's trailblazing Project Cyclops and the steerable Green Bank Telescope, Big Ear was pedestrian. As a passive telescope, it depended on Earth's rotation to survey the sky, so its choice of targets at any given moment was limited.[36]

The Ohio State search had the virtue of costing little, relying on volunteers, mostly undergraduate and graduate students, as well as several astronomy professors. One of these professors, Jerry Ehrman, noticed the famous "Wow!" signal, the closest SETI has come to succeeding. He had worked on the sky map and left Big Ear when funds ran out but returned as a SETI volunteer. One day in August 1977, Ehrman was flipping through hundreds of pages of printouts and saw a signal that came in on August 15 at 11:16 p.m. He noted the characteristics of a likely signal—strong, narrowband, and intermittent. Concentrated on only one of the fifty channels being monitored, the seventy-two-second signal was the kind whose origin should not have been natural. Ehrman wrote "Wow!" on the printout. The staff looked for weeks, but because Big Ear was a passive telescope it could focus on the source of the "Wow!" signal for only several minutes each day. Big Ear never again detected the "Wow!" signal.[37]

The "Wow!" signal remains a mystery. Perhaps, extraterrestrials transmitted again when Big Ear was focused elsewhere and, never receiving a reply, gave up. Perhaps the signal was leakage, not meant for interstellar communication. ET was not calling in the first place. It may be that the signal was natural and one day will be explained. Finally, the signal's origin may be human. During Project Ozma, Drake had detected a signal from a secret American aircraft. The "Wow!" signal may have come from another secret government project or sent by the notoriously secretive Soviets. Kraus has speculated that a human space probe beyond his security clearance sent the "Wow!" signal.[38]

NASA's SETI Workshops

Ohio State's SETI search was grassroots initiative, the world's first continuous SETI program, but Drake, Morrison, and Billingham had something grander in mind. "Big science" had defined their professional careers. Morrison had worked on the Manhattan Project during World War II; Billingham was a NASA administrator during the glory days of the Apollo missions; in the late 1960s, Drake was directing the Arecibo Observatory. All three envisioned SETI as a major government project. Yet by 1975 the Project Cyclops fiasco was reducing SETI's chances of NASA approval.

The time had come for Billingham to resell SETI within NASA. He realized that his first task was gaining "the support of the scientific community." Without that, SETI was "dead in the water." He assembled six SETI workshops, which Morrison consented to chair. Billingham later described these workshops as "the best decision" he "ever made."[39]

The workshops met from January 1975 to June 1976 under the auspices of NASA's Ames Research Center.[40] Sixteen leading scientists, in tandem with NASA people from Ames, scientists from Caltech's Jet Propulsion Laboratory in Pasadena, and others from NASA headquarters, discussed SETI. Their favorable assessment was no surprise, for Billingham had stacked the deck with SETI backers. Morrison, Drake, Sagan, Oliver, Bracewell, Greenstein, and A. G. W. Cameron were among the sixteen scientists running workshops. Also participating was Joshua Lederberg (1925–2008), biologist and Nobel laureate in medicine. Lederberg was very much interested in possible alien life, even the lowliest bacteria. Another participant was Charles Townes, a veteran of the 1971 Byurakan II conference. The rest were mostly astronomy professors.

The workshops broke no new ground. Their purpose was to repeat and validate the SETI narrative, but NASA published their report, thereby adding government approval of sorts to the speculations of SETI true believers. As could be expected, the workshops pointed out that scientists could learn much from extraterrestrials. The workshops also took the long view. It assumed older extraterrestrial races had faced many travails similar to those of Earth's twentieth century. With luck, the trajectories of older civilizations might give humanity deep insights into the direction of its own history, perhaps regarding the very meaning of the human journey, and whether and how evolution ends.

The workshops carefully projected a big-tent appeal. They noted that engineers and scientists would not be the sole beneficiaries of extraterrestrial knowledge. A wide range of professionals. "Anthropologists, artists, lawyers, politicians, philosophers, theologians"—in fact, "all thoughtful persons, whether specialists or not," stood to gain.

The workshops defined SETI as a win-win undertaking, for even failed searches carry a positive outcome. A universe found empty of intelligent life would give humanity "a strengthened belief in [its] near uniqueness." Besides, technology developed for SETI had other applications. As for potential dangers, SETI telescopes would listen but not transmit. Earth can choose to ignore an offensive message; the other civilization will never know of humanity's existence. The dangers are few, but the benefits are many, the workshops declared.

As always, the bottom line was money. Would a SETI program be expensive? With NASA budget hawks in mind, the workshops' answer was a simple no. Existing radio telescopes equipped with "low-cost state-of-the-art receiving and data processing devices" could do the search. For example, the spectrum analyzers of Project Cyclops could be replaced by fast Fourier analyzers, faster and cheaper, capable of tuning in on a million channels in real time. This new technology would suffice in the short run. Only afterward, if deemed necessary, should NASA build a dedicated SETI telescope. In short, NASA should start small and defer a major SETI investment for future consideration. Here was still another sign that NASA's SETI lobby had abandoned the grandiosity of a Project Cyclops, at least for the near future.

To emphasize the search aspects of the program, the workshop adopted the new acronym SETI (Search for Extraterrestrial Intelligence), to replace CETI, (Communication with Extraterrestrial Intelligence), the old name, introduced in 1965. "Search" implies exploration, whereas the term "communication" carries a whiff of science fiction, a toxic association when trying to get respect. The name change also differentiated American from Soviet SETI, since Soviet astronomers continued with the term CETI.

The workshops concluded that SETI's rightful home was in NASA because SETI and NASA shared the same mission of exploration. In 1976, Billingham was put in charge of the SETI Program Office at the Ames Research Center. His title was Chief of the Extraterrestrial Research Division. According to Drake, this made Billingham "the first person in the United States to head a civil service unit giving official recognition to alien civilizations."[41] Meanwhile, Bruce Murray (1931–2013), the new director of the Jet Propulsion Laboratory, was also supporting SETI. In 1978, Congress appropriated $2 million to NASA for SETI research. SETI had seemingly arrived—no longer dismissed as a kook pursuit of scientists who had read too much science fiction, no longer ufology's close relative. However, life is fraught with friction. SETI had to pay the price of success. At the same time that Billingham was normalizing SETI within NASA, a reaction against SETI was setting in.

8

SETI in NASA

Rise and Fall

Every scholar dreams of making an impact, of producing insights that are debated and hailed, even assailed, but never ignored. After the sound and the fury, the scholar's peers may or may not have reached a consensus, but this much is clear: if the old narrative has been radically changed, revised, or at the very least called into question, the scholar has realized the dream; he or she has made an impact.

Michael Hart Restates the Fermi Paradox

In the story of SETI, Michael H. Hart (b. 1932) of Trinity University made this impact. His 1975 article in the *Quarterly Journal of the Royal Astronomical Society* challenged the SETI belief that our galaxy is full of intelligent life.[1] He challenged by asking the simple question why this life has not been seen. Surely, Hart argued, at least one of the thousands of advanced extraterrestrial races, which SETI advocates were taking for granted, should have visited Earth. Hart was echoing the village atheists of old who confounded the pious by asking why God has not been seen.

Hart rejected the traditional SETI scenario, which claims that the vast expanse of space and the speed-of-light barrier prevent long-distance space flight. Hart presented an alternative scenario in which extraterrestrials had no need to travel directly from their world to Earth.

Instead, they leapfrog from planet to planet over generations—like South Sea Islanders settling the remote islands of the Pacific. In Hart's universe, an extraterrestrial civilization explores over great distances through colonization. A science-fiction gimmick like *Star Trek*'s warp speed is not necessary.

That extraterrestrials are unknown and unseen, Hart called Fact A. He gave four possible explanations for Fact A, each of which he sought to demolish. He began with physical explanations, such as the enormous "distances between the stars" and the enormous energy requirements to travel these distances. Yet, Hart argued, distance is no obstacle if extraterrestrials lived longer, traveled in an induced sleep, or sent robots. As for energy, Hart noted the theoretical work on alternative technologies.

Most of the proposed explanations for Fact A, according to Hart, are sociological: extraterrestrials avoid exploration, they have destroyed themselves, or they are deliberately ignoring Earth. Yet, Hart asked, are these sociological explanations true for all extraterrestrials at every stage in their history? At one point, at least one alien civilization should have traveled the stars. Besides, said Hart, humans are curious, energetic, and adventurous. We know nothing about extraterrestrials, yet it is said they have not arrived because they are lazy, self-absorbed isolationists, behaving very differently from humans. Hart was throwing the mediocrity principle back at SETI activists.

Another category of explanations for Fact A is the temporal, which maintains that extraterrestrial civilizations have evolved only recently and are not advanced enough to visit us. Believing this, said Hart, is to believe humanity is at or near the vanguard of galactic progress. Hart left unmentioned the implication that extraterrestrials near the human level will teach us little. So much for the cancer cure, immortality, and the recipe for world government!

As for the possibility of past extraterrestrial visitations, Hart argued that the SETI community had to show why extraterrestrials visited rarely, if at all, and why they did not colonize. In a galaxy full of advanced life-forms, at least a few must have had the wanderlust and the urge to settle down. Here, Hart was treading on shaky ground. After rejecting sociological explanations for extraterrestrial absence, he was asking SETI to use these very explanations. Finally, Hart mentioned and dismissed the UFO hypothesis: extraterrestrials are here but the political/scientific establishment denies their presence. Hart saw no reason to discuss the UFO hypothesis, since few astronomers accept it.

Hart's conclusion was unsparing: extraterrestrials have not arrived because they do not exist or they are few. Suddenly, the SETI world was on the defensive, having to explain the absence of contact—why just one of the thousands of alleged alien civilizations has not revealed itself, either by coming to Earth or at the minimum by sending a radio signal.

Hart may not have been the first to reach this conclusion. In 1950, Enrico Fermi had asked why extraterrestrials had not been seen. Probably the most respected physicist of his day, Fermi had conducted the first nuclear chain reaction in 1942. Working at Los Alamos after World War II, he and his companions were having lunch one day and were discussing the alleged flying-saucer sightings, a hot item in the news. They were wondering whether flying saucers could exceed the speed of light. After the conversation had veered to other topics, Fermi suddenly returned to subject of extraterrestrials and asked, "Where is everybody?"[2]

What was Fermi thinking when he asked this question? Was he rejecting the idea that extraterrestrials exist? The belief that this was his thinking has given rise to an alleged truism, the Fermi Paradox, which holds the lack of contact as inconsistent with extraterrestrial existence. However, in a second version of Fermi's state of mind, there is no paradox. Fermi was merely stating that the speed-of-light barrier prevents interstellar travel. Whatever may have been Fermi's thoughts, Hart made the so-called paradox a topic of debate. For this reason, Shklovskii said that the paradox should bear Hart's name instead of Fermi's.[3]

At first the SETI establishment tried to ignore Hart, who later claimed that criticizing SETI was easier in British scientific journals than in American ones.[4] After Hart had published in Great Britain, he could no longer be ignored. He had struck a raw nerve.

The International Astronomical Union's 1979 Assembly

When the International Astronomical Union (IAU) met in 1979, both Hart and SETI were on the agenda. A special one-day session focused on "Strategies of the Search for Life in the Universe." This attention was a coup for Hart, and for SETI it was another sign that it had arrived in astronomy's high circles. Although Sagan was absent, SETI was well represented, with Drake, Morrison, and Oliver as well as Shklovskii, Kardashev, and Troitskii.

The IAU represented astronomers from nearly fifty countries. Founded in 1919 and based in Paris, it met every three years at a different venue. In 1979, Montreal was the site of its Seventeenth General Assembly.[5]

The morning session discussed the number of technological civilizations in the galaxy, the N in the Drake Equation. Michael Hart spoke first and mostly repeated his 1975 paper: the absence of an extraterrestrial presence on Earth implies a sparsely populated galaxy. Hart did add a new twist; he gave a reason for N's small value. Like many biologists, he claimed that "the spontaneous origin of life" was very improbable on Earth, a one-in-a-billion event, and that believing in an abundant galactic life is wishful thinking.

In his presentation, Drake probably realized that huge estimates for N energized skeptics like Hart. Drake prudently offered a smaller value for N, about 100,000 intelligent civilizations, a figure that reduced the likelihood of ET's arrival but was not small enough to demolish Hart's colonization thesis.

Drake argued best when he pitched the idea that extraterrestrials had managed overpopulation with a simpler solution than colonization. They had achieved zero population growth, making galactic colonization unnecessary. Drake was alluding to the population pressures that historically drove mass migrations on Earth. To apply this engine to extraterrestrials seemed logical in the postwar years when high birth rates had awakened Malthusian pessimists. This was one reason why Hart struck a nerve. Unfortunately for Hart's thesis, birth rates have since declined, and demographers predict a smaller world population by 2100. Hart's colonization scenario carries less weight today. In our presentist age, Drake's counterthesis of zero population growth in the galaxy is more compelling than ever. Drake did not throw a knockout punch, but if sports metaphors can be mixed, he did hit a home run.

When Bernard Oliver took the floor, he could be expected to speak his mind. Oliver denounced SETI's new opposition as "peculiar" for giving "too much credence to that fantastic extrapolation, the Kardashev Type II civilization." Unable to find solid evidence for the astroengineering of this supercivilization or for an alien visit, this "peculiar opposition" had resorted to silliness, insisting that advanced extraterrestrials are either quarantining Earth (the Zoo hypothesis) or are few or nonexistent. All this speculation, Oliver intoned, ignores the real world of energy. A rocket traveling at 70 percent of the speed of light requires energy sufficient "to power the United States with electricity for over 100,000 years." Extensive interstellar travel is an illusion, and Oliver named the villain. It was science fiction, which had "brainwashed" the public and scientific community. Oliver was telling Hart that space is too vast even for colonization.[6]

In spite of Oliver's fulminations, Sebastian von Hoerner echoed science fiction when he focused on interstellar travel. Energy to power spaceships is at hand, said von Hoerner; it is found in the stockpile of the world's atomic bombs, which in 1972 contained energy equivalent to 40,000 megatons of TNT. Von Hoerner recalled Project Orion of the 1950s and 1960s, which designed an interstellar spaceship propelled "by exploding nuclear bombs far behind, [and] catching the energy with a large shield and a long shock absorber." Present technology can build this ship, he reminded the audience.

Having implied that extraterrestrials can zip through space, von Hoerner had to account for their non-arrival. He gave the usual excuses, even conceding that humanity "may actually be alone," but he gave the most attention to self-destruction: Earth's "frightening arms race" had "completely gotten out of hand," he said, and "cannot continue for long." To claim that all advanced extraterrestrials have self-destructed said more about von Hoerner's pessimism than it did of ET's fate.

A popular SETI excuse for non-contact has been that the searches have been very sporadic. Benjamin Zuckerman (b. 1943) and Jill Tarter

(b. 1942) of UC Berkeley provided details at the conference. Not only were there "fewer than 20 radio searches," but SETI astronomers were using "equipment originally designed and constructed for radioastronomical observations." No SETI search that Zuckerman and Tarter knew of used dedicated hardware able to catch the "leakage signals" that extraterrestrials were probably sending. Without this hardware, examinations of nearby stars would miss emissions similar to Earth's TV carrier signals and strong military radars.

That SETI had barely peeked into the vast expanses of space might explain the absence of radio contact. This argument, though, does not face Hart's colonization thesis. Rebutting Hart requires showing why an advanced civilization has not arrived. Drake had already pointed to zero population growth. Philip Morrison offered still another explanation: extraterrestrial societies realize the limits of growth. Intergalactic Greens frown on waste and keep their societies approximately at Earth's level. Morrison was implying that technological progress reaches a standstill and that Type II and Type III supercivilizations do not exist. As for Earth-like Type I civilizations, they do not have the capacity to colonize. All that remains is for Earth to search and communicate. Whereas Hart put the burden of contact on ET, Morrison put it back on Earth. In what was the great SETI tautology, Morrison repeated the exhortation of his seminal 1959 article but in language that reflected the pessimism of the 1970s. He preached the message that SETI true believers wanted to hear: ignore Hart and keep searching. Like Drake's zero population growth, Morrison's extraterrestrial Green sensibility catered to the prevailing mood in smart circles, and for this reason it was a home run.

What did Hart achieve? He made astronomers think twice about Type II and Type III civilizations.[7] If these civilizations were fantasies, so was the basket of goodies they would bestow: immortality and solutions to whatever ailed Earth. If Hart aimed to kill SETI, he failed. His real success was in launching a debate that has continued to the present, mirroring the age-old debates between people of faith and skeptics. SETI has spawned its version of atheists, agnostics, and believers.[8]

Frank Tipler and the Von Neumann Machines

What Hart had hinted, Frank Tipler (b. 1947) of UC Berkeley boldly proclaimed in his 1981 article, "Extraterrestrial Intelligent Beings Do Not Exist." Tipler revived Bracewell's probe theory but reworked it to deny the existence of extraterrestrials. Tipler visualized a near future in which Earth's rocketry and computer technology will produce self-replicating probes. These he dubbed "von Neumann machines" after the Hungarian-American mathematician John von Neumann (1903–57) who had predicted their development. If humanity is typical, as the SETI community insists, nothing less should be expected from advanced extraterrestrials, Tipler said. Alien von Neumann machines should be busy exploring the galaxy, mining raw materials for repairs along the way, even reproducing themselves. Having lives of their own, von Neumann machines will outlive their creators. In this manner, Tipler drastically reduced the cost of interstellar exploration and made the arrival of nearly immortal von Neumann machines almost seem probable.

Tipler added that a third of our galaxy's stars predate the sun. Their planets should have produced life long before Earth. Even if none of these civilizations has colonized, at least one should have sent out probes. The principle of mediocrity demands a curiosity similar to humanity's. Tipler's conclusion was uncompromising. Von Neumann machines have not been detected, because they and their creators do not exist.

Tipler did not stop with ET. He rejected SETI itself, dismissing the whole thing as wishful thinking, the scientific version of ufology, seeking succor, even a salvation of sorts, from the stars. Tipler named names: Sagan, Hoyle, Cameron, and Drake, he claimed, were especially prone to replacing traditional gods with new ones. Tipler was an atheist in religion and likewise with SETI. If nothing else, he was consistent.

Tipler got attention. Sagan gave a politically correct rebuttal: he could not see ethically advanced aliens strip mining their way through space. On the other hand, NASA scientist John Wolfe opined that the resource-sucking von Neumann machines threatened other civilizations who promptly destroyed them. Biologist Stephen Jay Gould was thoughtful. Claiming to be baffled by the thoughts of humans in different cultures, he found the workings of the extraterrestrial mind a greater mystery.[9]

Tipler's parting shot was insisting that the intellectual climate had turned against SETI. He pointed out that Shklovskii no longer believed the galaxy contained intelligent life and that Freeman Dyson was also doubtful.[10] Defections aside, Tipler was wrong about SETI's alleged declension. Similar to a new church, the SETI community was merely losing followers who had second thoughts. SETI was not declining, though. In fact, it was gaining respect—from the top of NASA, no less.

SETI Gains Respect

In June 1979, NASA's Ames Research Center sponsored a two-day conference on the theme "Life in the Universe." Robert A. Frosch (b. 1928), NASA administrator, gave the keynote address and outlined a new mission for SETI, in effect giving it the unofficial approval of NASA. Frosch saw a successful search transcending humanity's immediate welfare, for it would reveal that humanity has "siblings." In short, the aim of SETI was to clarify humanity's relationship to a living universe.[11] Although Frosch and the SETI community would deny it, this mission has semi-religious overtones, for classifying humans according to beings beyond Earth replaces the traditional hierarchy of angels above and beasts below.

In 1981, with a new president (Ronald Reagan) in the White House, Frosch was no longer running NASA, but SETI continued to gain respect—in this case from the International Astronomical Union. At its previous meeting in Montreal, SETI's legitimacy had been questioned. In 1982, at its Eighteenth General Assembly in Patras, Greece, the IAU formally recognized SETI by establishing Commission 51—Search for Extraterrestrial Life. One of Commission 51's tasks was searching for alien radio signals. Michael Papagiannis (1933–98), professor of astronomy at Boston University and a SETI supporter, was president of Commission 51; Drake and Kardashev were vice-presidents. Commission 51 grew rapidly and within three years numbered over 250 members from more than twenty-five countries.[12]

In 1984, the IAU met in Boston and Commission 51 was renamed Bioastronomy: The Search for Extraterrestrial Life. SETI had been subsumed into something bigger, bioastronomy, the search for all alien life, including the microscopic.[13] The search for microscopic life was very respectable, the main activity, in fact, of NASA's Viking expedition to Mars. Putting SETI into such company was normalizing it.

Still another sign of professional favor came from the National Academy of Sciences. Every ten years, the NAS identified the priorities for astronomy. Whereas the Greenstein Report of 1972 had lauded SETI but did not call for funding, 1982's Field Report, named after its principal author, George B. Field (b. 1929), recommended that SETI receive $20 million over the decade, declaring the time "ripe for initiating a modest [SETI] program."[14]

The Great Silence, or Why ET Has Not Called

Yet ET remained embarrassingly silent. In SETI's early days, contact had been seen as the simple matter of tuning in on the proper intergalactic radio channel. In the 1980s, although searches had been few, it was obvious that extraterrestrials were not bombarding Earth with radio signals, eager to begin a conversation. Blaming failure on not locating ET's radio frequency was sounding stale. More than ever, other explanations for failure were necessary.

In 1983, David Brin (b. 1950), a space scientist noted for writing science fiction and consulting, wrote "The Mystery of the Great Silence," which asked why extraterrestrials, if they exist, have not contacted Earth. Brin grouped the many explanations for the "Great Silence" into eight major categories. First was the rarity of intelligent life. If intelligence is instead common, the "graduation" factor, the second category, may explain the Great Silence: advanced galactic races have outgrown radio, communicating in ways beyond human technology. The third category is "timidity." ET does not care to travel or communicate. The fourth category is the opposite of timidity: extraterrestrials are star trekkers yet avoid Earth, judging humans dangerous, dull, or immature. In the fifth category, extraterrestrials once came to Earth, were bored, and moved on. The sixth category is "frightening": it may be that the universe has "dangerous natural forces." Planets end up like Mars and Venus or are prone to space junk like asteroids. The seventh category includes "dangerous unnatural forces," or the misery that intelligence produces. According to this scenario, extraterrestrials eventually self-destruct, or paranoid species deliberately destroy them. Earth has been lucky so far. Finally, Brin presented the "water worlds" category. Many planets have vast oceans with intelligent life that lives in water. The universe may be full of intelligent crabs who exchange love poetry and martial ballads but are clueless about technical communication.[15]

The verdict on Brin's "Great Silence"? The usual SETI apologia applies. SETI has been an outlier. Forced to compete with traditional astronomy for telescope time, its searches have been few and brief; devised for other purposes, not even its equipment is dedicated. These excuses for failure SETI true believers could live with; unfortunately, they had to sell their optimism to the United States Congress.

Congress and NASA SETI

Congressional approval of a SETI program in NASA was crucial, for only NASA could afford expensive hardware. After the Project Cyclops fiasco, Billingham pursued a patient incremental policy for SETI. He used his position as chief of NASA's Extraterrestrial Research Division to fund a few small SETI studies and projects.[16] In 1978, NASA SETI had received an appropriation of $2 million, a trifling sum, but enough to be noticed. If politicians deemed SETI a waste, they could defund NASA's SETI program and set back the search.

In 1978, Senator William Proxmire (1915–2005), a Democrat from Wisconsin, launched his crusade against SETI. A self-appointed watchdog over government waste, Proxmire had a genius for publicity and a knack for concrete images. He invented a Golden Fleece Award to focus attention on federal spending he deemed frivolous. One Golden Fleece Award "honored" the National Science Foundation for studying why people fall in love; another award honored the study of a Peruvian brothel, often visited by the researchers "in the interests of accuracy."[17]

On February 16, 1978, Senator Proxmire called a press conference to announce that SETI was receiving a Golden Fleece Award. Proxmire blamed the *Star Wars* movie mania then sweeping the nation for seducing NASA. Proxmire was making NASA seem ridiculous, and no one wishes to appear ridiculous. This may explain why after Proxmire spoke, the joint House and Senate Appropriations Committee canceled SETI funding for fiscal year 1979, although the full House and Senate had authorized $2 million the previous year.[18]

SETI did have congressional supporters, however. On September 19–20, 1978, the subcommittee on Space Science and Applications of the House Committee on Science and Technology held the first hearings on SETI.[19] The subcommittee stacked the witness list in SETI's favor. Among those who testified were Morrison, Oliver, Berendzen, Cameron, Noel Hinners (a NASA administrator), David Heeschen (director of the Green Bank Observatory), and George Pimentel (deputy director of the National Science Foundation). Only Pimentel was guarded about SETI; he testified that astronomers should do their normal work and hope for serendipity. Pimentel did call SETI "worthwhile," although he doubted whether contact would occur during his lifetime. Pimentel was right; he died in 1989.

The subcommittee chairman, Representative Don Fuqua (b. 1933), a Democrat from Florida, was seeking to help NASA SETI. Clearly, he was referring to Proxmire when he asked the witnesses for the best reply to the senator of the "famous award." Although Fuqua wanted ammunition, he got none; no magic bullet existed. The SETI people could not guarantee contact. Berendzen and Morrison defined SETI as cutting-edge research with spillovers. They saw a need to educate the public, a truism unfortunately similar to a politician's excuse after losing an election. These justifications would not silence Proxmire, who could easily reply that SETI funding was better allocated elsewhere. Morrison did make a clever point. Eighty-five percent of the public, he claimed, probably believed that SETI was "the purpose of much astronomical work." He was implying popular approval for a NASA-sponsored SETI search.

Representative James Lloyd (1922–2012), a Democrat from California, a retired navy pilot, perhaps made the best case for SETI. Lloyd told the subcommittee he had read science fiction, watched it on television, and loved *Star Wars*. The SETI testimony had left him "spellbound."

Lloyd was implying that SETI was born from the heart, not the mind, born of the same urge that made people gamble and walk into strange places. Unlike Apollo moon trips, SETI was not meant to be scripted from beginning to end. As the SETI people themselves said, the SETI quest recalled Europe's Age of Discovery, when explorers beginning with Columbus often relied on a gut feeling, surprising themselves and the world with their findings.

Fuqua's hearings did not help SETI. To exclude ufologists, the hearings were little publicized and as a result sparsely attended. Media coverage was likewise sparse. Well-known SETI personalities like Drake and Sagan were absent. Sagan would have guaranteed attention, now that his television appearances had made him a media superstar, well known beyond the narrow circles of science and academia. The final blow was world events. On September 17, Egypt's Anwar Sadat and Israel's Menachem Begin signed the Camp David peace accords. With history being made, events leading to Nobel Peace Prizes, media and congressional attention was elsewhere.

NASA's SETI funding was not restored for the 1979 fiscal year. All the same, NASA SETI did not die. NASA was far from conducting SETI searches, still in the research stage of determining strategy and hardware. Billingham found unspent money and placed SETI in NASA's exobiology division, while waiting for the Proxmire storm to end. Proxmire found out about Billingham's ploy, and the storm turned into a hurricane.[20]

On July 30, 1981, Proxmire had a busy day on the Senate floor. He started off suffering a defeat on NASA's budget. He wanted to reduce it, especially since Congress was willing to cut social programs. The Senate instead increased funding for NASA's research and development, even beyond President Reagan's recommendation. Proxmire had better luck with SETI. He proposed the following amendment (no. 338) to the NASA appropriation: "That none of these funds shall be used to support the definition and development of techniques to analyze extraterrestrial radio signals for patterns that may be generated by intelligent sources."

In support, Proxmire claimed that Congress had clearly intended in 1978 to end NASA SETI, but NASA had dishonestly continued the

program. He wanted to pull the plug before the SETI program bal-
looned and ended up costing the nation at least $50 million over a
ten-year period. Proxmire next attacked SETI itself, citing Frank Tipler
for the nonexistence of advanced extraterrestrials. Dismissing SETI as
a "ridiculous waste of the taxpayer's dollars," Proxmire was giving the
Senate a chance to show financial restraint, after it had increased the
NASA budget. His SETI amendment passed.[21]

Frank Drake quickly reacted. He called Proxmire's office for a copy
of the Golden Fleece Award, only to be told that no such paper existed;
the award was rhetorical. Drake did not let up; he nominated Proxmire
to membership in the Flat Earth Society. More seriously, he accused
Proxmire of hypocrisy, opposing SETI to save money yet supporting
subsidies to the dairy industry, a key interest group in his home state
of Wisconsin.[22]

Proxmire's crusade revealed that SETI needed better public relations.
Sagan went to work collecting names, and in a letter published in the
October 29, 1982, issue of *Science,* sixty-nine scientists expressed their
support for SETI. The letter was cleverly worded, referring to unfet-
tered scientific inquiry, so that even lukewarm SETI supporters could
sign it. Apart from the SETI grandees, signers included scientists not
particularly associated with SETI, such as Francis Crick, Stephen Jay
Gould, Linus Pauling, and Stephen Hawking. Even Shklovskii signed it,
although at this point he had become a SETI agnostic if not an atheist.

Meanwhile, NASA did not panic. Hans Mark, NASA's deputy administrator, called a meeting of top NASA officials, and they decided
to write two budgets. The first obeyed Congress and terminated NASA
SETI for 1982. The second was a 1983 budget that restored funding in
case the storm ended. The next task was lobbying Congress to have that
funding restored.

The key was Senator Proxmire. He was informed of the Field Report's approval of SETI and was personally lobbied. Billingham later
recalled many lines of communication to the senator. The best known
was the personal appeal from Carl Sagan, then riding high as America's
superstar scientist, winner of the 1978 Pulitzer Prize in general nonfic-
tion. Sagan told Proxmire that through contact humanity might secure
advice on avoiding nuclear war. Proxmire was impressed and promised

to leave SETI alone, although he never publicly disowned his previous hostility.[23]

Although NASA SETI survived, a potential danger had been revealed. As the SETI magazine *Cosmic Search* noted, public support for SETI was not evident during the Proxmire wars.[24] The underlying reason was not mentioned. Ufology invariably contaminates public discussion of extraterrestrials. It should be recalled that concern over ufologists kept the Fuqua subcommittee from publicizing its SETI hearings. The SETI lobby had to be exclusive. When it lobbied Proxmire, it did so behind closed doors. Elites were interacting. The future of NASA SETI, pro or con, would not be determined by grassroots pressures.

After Proxmire: NASA SETI Resumes

Proxmire's silence allowed NASA SETI to receive $1.65 million for fiscal year 1983, freeing it from the research closet. However, the time spent in the shadows had not been wasted. In 1979, NASA made a major decision on SETI strategy, a political compromise that ended an internal turf battle between two of its branches, the Ames Research Center, which advocated targeted searches, and the Jet Propulsion Laboratory, which preferred all-sky searches.

How do targeted and all-sky searches differ? A targeted search, such as Project Ozma, examines individual stars similar to the sun for signals. This strategy hopes extraterrestrials from the observed star system are beaming radio signals at the same time the radio telescope is scanning. If a planet from another star system is beaming, however, Earth is tuned to the wrong place. Hence, the all-sky survey advocates, led by Bruce Murray, JPL's director, insisted that scanning a large patch of sky on a wide band of frequencies will more likely succeed. There is a drawback. Unlike a targeted search, this strategy will not catch a weak signal or eavesdrop on internal chatter, for it presupposes ET is beaming a very strong signal, probably meant for detection. Since either approach could succeed, NASA adopted both and kept the peace between its Ames and JPL branches. Ames, though, remained the lead center for NASA SETI.[25]

Although NASA SETI was receiving an annual appropriation of $1.5 to $2 million during the 1980s, NASA did not formally endorse SETI until 1988. Even then, sailing was not smooth, for new opposition to SETI emerged in Congress. In 1990, NASA requested $12 million in SETI funding for the next fiscal year. The House Appropriations Committee reduced the request by 50 percent. Then, Representative Silvio Conte, a Republican from Massachusetts, took aim at the remaining $6 million. Addressing the House in June, Conte exploited the "giggle factor" when he accused NASA of looking for "little green men with misshapen heads," wasting money when Americans could not find "affordable housing." He continued: "Of course, there are flying saucers and advanced civilizations in outer space. But we don't need to spend $6 million this year to find evidence of these rascally creatures. We only need 75 cents to buy a tabloid at the local supermarket." Conte regaled the House with tabloid headlines: "Space Aliens Stealing Our Frogs!" and "Noah's Ark Was Built by Space Aliens." The House voted to delete the $6 million SETI funding from NASA's 1991 budget.[26]

Nonetheless, NASA SETI survived. In the Senate, Barbara Mikulski of Maryland and Jake Garn of Utah saved it in 1990. Garn was a former astronaut who recalled: "When I was in orbit, I gazed out into space, and it was clear to me that there had to be life out there." It may have helped that Garn is a Mormon and that the Church of Jesus Christ of Latter-day Saints has a very open view toward extraterrestrial life.[27]

NASA SETI had to wait till fiscal year 1991 for the go-ahead to begin a ten-year, $100 million project. Only then did NASA construct the dedicated equipment, a multi-channel spectrum analyzer capable of scanning fifteen million channels simultaneously.[28] The system would identify a likely signal in real time so that a human operator immediately knew, instead of after the fact as in the case of the "Wow!" signal. Computer technology had automated the search.

The search officially started on October 12, 1992. The selection of Columbus Day 1992, was deliberate. The year 1992 marked the quincentennial of Columbus's voyage, an event that profoundly changed the course of history. The SETI people believed they too were about to reshape humanity, but far more than Columbus, who merely introduced

Europeans to other humans, and vice versa. Here was optimism as well as a trace of hubris.

The Ames search used the Arecibo radio telescope to focus on a specific star system for several minutes. It looked for narrowband signals, hoping an intelligence had reduced signal dispersion by narrowing the frequency range. Ames began with the nearer stars in case their planets' inhabitants were responding to Earth's first television signals. In all, Arecibo planned scanning seven hundred stars within eighty light-years of Earth.

The JPL all-sky search used the Goldstone Deep Space Communications Complex to survey frequencies from 1 to 10 GHz. By 1999 the Goldstone telescopes would have covered 320 million channels. Although the JPL search focused on a specific patch of sky for only a second, this brief moment was sufficient if extraterrestrials were transmitting a beacon.

The SETI community was excited. After years of planning and lobbying, beating back attacks from Congress and the academy, a full-time search had finally started. The combined Ames and JPL efforts had no precedent. NASA bragged that within the first half hour of operation the sum of all previous searches had been exceeded. An ecstatic Drake predicted success by the year 2000.[29]

With these expectations, it is not surprising that the SETI community devised contact protocols to verify and announce the reception of a signal. Billingham, along with Michael Michaud, the State Department's expert on space exploration, and Jill Tarter, who had joined NASA SETI in 1989, worked with the International Academy of Astronautics on protocols. In 1999 they issued the "Declaration of Principles Concerning Activities Following the Detection of Extraterrestrial Intelligences." The Declaration had two essentials: having other astronomers verify the signal to avoid embarrassment, and avoiding secrecy. Tarter said: "The signal isn't being sent to California. It's being sent to planet Earth, and everyone deserves to know." Wi hin three years, six international space societies adopted the Declaration.[30]

The Declaration has no legal standing; it is not a treaty, only a gentleman's agreement among the world's SETI searchers. It is a system of behavioral rules, typical of the countless arrangements that humans have devised throughout history to prevent anarchy, and which the selfish have broken. Nothing prevents an astronomer, when informed of the signal, from claiming credit for contact to become world famous. Whichever scenario plays out—the Declaration observed or broken— must await contact, if that event occurs.

The SETI Debate Again

If NASA SETI received an informative alien message, what would it tell us of ET? Carl Sagan in his SETI novel *Contact* wrote of nearly omnipotent aliens who share transcendental wisdom. Frank Drake in his 1992 SETI memoir repeated his belief in extraterrestrial immortality. Bernard Oliver had a different vision. Retired from Hewlett-Packard, he was now a full-time SETI person, head of NASA's Ames SETI office from 1983 to 1993. Oliver focused on contact's effect on humanity. He foresaw other nations erecting their own antennae, taking a first step in uniting humanity, a "spin-off" as valuable as contact itself. Oliver was harking back to the world government dreams of the 1950s.

Skeptics were legion, and among them was Ernst Mayr (1904–2005), evolutionary biologist at Harvard, who once called SETI "naïve" for believing that extraterrestrials were using Earth's twentieth-century radio technology. In March 1993, Mayr published a letter in *Science* in which he denied that intelligence conferred an evolutionary advantage. Of the fifty billion species in Earth's history, he noted, only one achieved intelligence. Defining SETI as "biological and sociological," Mayr was astounded that NASA allowed astronomers and engineers to approve it without a "more broad-based consultation," meaning of course the presence of biologists.[31]

Besides the skepticism many biologists harbor toward advanced extraterrestrial life, Mayr was doing something common in academia: he was defending his turf. Physical scientists were presuming to know more of evolution than biologists. Mayr was taking umbrage. Academic territoriality should not be dismissed, though, as a variety of modern atavism. More than other academics, scientists need funding.

Often caught in a zero-sum game, they easily see waste among their competitors.

In his response to Mayr, also published in *Science*, Frank Drake insisted that humans were simply the first terrestrials to get smart. If not *Homo sapiens*, another species would have developed intelligence. Drake chided Mayr for underestimating "the opportunistic nature of biological systems or the enormity of cosmic time." Drake was referring to his 1981 article about Stenonychosaurus (now called Troodon), a dinosaur that paleontologist Dale Russell (b. 1937) discovered. This creature walked upright, surveyed its world with large eyes in front of its skull, and sported a large, opposable thumb. Sharing the same brain size of some modern mammals, it was on "the verge of becoming intelligent." In Drake's opinion, this dinosaur showed that evolution, despite twists and turns, has directionality and that intelligence was no fluke. Troodon has heartened SETI circles, which see this creature's descendants filling the intelligence niche, if not for the Cretaceous extinction.[32]

There were astronomers who shared Mayr's skepticism. Gerrit Verschuur had been a pioneer SETI researcher. He later turned against SETI, arguing like Mayr that the contingency and unknowns in evolution make a meaningful conversation with extraterrestrials unlikely. In Verschuur's opinion, hoping that extraterrestrial advice will save the planet showed that SETI is "a technological search for God." Sebastian von Hoerner, also once identified with SETI, likewise joined the apostates, as Shklovskii had done before his death in 1985.[33]

Senator Richard Bryan Ends NASA SETI

With NASA SETI having congressional enemies, a staffer from Senator Mikulski's office advised NASA to change the SETI program's name and lower its visibility.[34] The advice was taken. The "SETI Microwave Observing Project" was reborn as the "High Resolution Microwave Survey" and, with the telltale word "SETI" dropped, moved from NASA's Life Sciences Division to the Solar System Exploration Division. The name change was a badly kept secret. The *New York Times* (October

6, 1992) reported it along with NASA's denial of a disguise. According to NASA, the new name only reflected the program's wider objectives: the very real possibility of other astronomical discoveries. Not to be discounted, NASA continued, was the new signal-processing technology, which could be applied to "telecommunications, air-traffic control and seismology." In brief, NASA SETI had outgrown the detection of alien signals.

Senator Richard Bryan (b. 1937) disagreed. In addressing the Senate on September 21–22, 1993, the Nevada Democrat insisted that detecting alien radio signals remained the real mission of NASA's SETI program. Its name had been changed and it had been "buried deep in the NASA bureaucracy" only to trick Congress. Bryan carefully noted that he was not denying SETI's scientific merits. Unlike Silvio Conte, who had died in February 1991, Bryan distinguished between SETI and ufology. Bryan professed outrage less at SETI itself than at NASA, which he accused of being out of control and willfully ignoring the wishes of Congress. By making a deceitful bureaucracy his target, Bryan cleverly reinvented the case against NASA SETI.[35]

NASA SETI had the bad luck of being caught in a perfect storm. Apart from the perennial "giggle factor," the program was too small for its termination to affect a major contractor. Furthermore, Congress was having a spasm of deficit guilt. Budget cutting was the rage, and Bryan was playing deficit hawk who preferred butter over stars, who told Congress the $12.3 million NASA wanted for SETI could buy 135 homes for needy families in Nevada or provide day care for 3,400 toddlers so that single parents could find work. Also in Bryan's favor was his timing. He was attacking NASA when its reputation was low, suffering the recent fiasco of the Hubble Space Telescope, which sent into orbit amid much fanfare had failed to focus properly. Revelations of massive contractor fraud also darkened NASA's image. Bryan's charge of bureaucratic deceit was the final straw.[36]

Senator Garn had retired, leaving others to rebut. Senator Mikulski offered classic SETI apologia when she described SETI as "serious science with serious applications," which would answer some of the most serious questions "asked for thousands of years." Senator Phil Gramm (b. 1932) of Texas got closer to the substance of Bryan's brief when he noted the very modest cost of the program as compared to its potential gains.

Unfortunately for NASA SETI, the public was not involved.[37] Congress voted primarily on its perception of SETI. According to Verschuur, his letter criticizing SETI had circulated in Congress.[38] Bryan pressed the right buttons in replying to Mikulski and Gramm. He again mentioned the deficit, the need for priorities, and the alleged circumventing of Congress's wishes, all contributing to the public's "skepticism" and "cynicism." No one likes being tricked, and Bryan insisted Congress had been tricked. Bryan tacked an amendment to kill NASA SETI on NASA's 1994 appropriation bill. On September 22, 1993, the Senate approved the amendment by the wide margin of 77 to 23. This decision was confirmed on October 1, when a House-Senate conference committee eliminated SETI funding from the NASA budget.

Bryan was relentless. He and his staff refused to talk to SETI's supporters. To prevent the next session of Congress from restoring funding, $1 million was appropriated to dispose of the equipment and wipe out NASA SETI.[39] In November, Bryan informed NASA that he was watching to make sure it was obeying Congress's wishes.[40]

Although Drake, Tarter, Oliver, Billingham, in fact, all the SETI people, were shocked and saddened, to say the least, the termination was inevitable. Often in congressional crosshairs, NASA SETI was destined to suffer a fatal wound, if not in 1993, then later. NASA SETI could never have overcome its great liability: its inability to guarantee success. In a speech of September 1992, Oliver confessed that NASA's planned ten-year search might be insufficient, making an open-ended commitment necessary, something that was not the NASA way.[41] Although destined for the dust heap, NASA SETI cannot be dismissed as a failure. Its very existence represented a government seal of approval, a legitimacy of sorts for SETI, which in its early years had struggled for respect.

Non-NASA SETI

With NASA out of the picture, the shoestring searches, both American and foreign, once seen as minor, which the big NASA operation would

eclipse, now represented the SETI future. In 1993, Tarter and Billingham placed sixty-one shoestring searches into three categories. The
first category contained brief searches such as Project Ozma. Next was
the reanalysis of astronomical data gathered in non-SETI research.[42]

The best example of data reanalysis was SERENDIP, an acronym
for the Search for Extraterrestrial Radio Emissions from Nearby Developed Intelligent Populations. SERENDIP was born at UC Berkeley,
the brainchild of Dan Werthimer (b. 1954) and Stuart Bowyer (b. 1934),
who installed the hardware and processing software at the Hat Creek
Observatory in 1980. It still running today. SERENDIP dealt with a
major problem in SETI: its limited access to radio telescopes. Werthimer and Bowyer devised an ingenious solution: "piggybacking" on
non-SETI projects, thus recording frequencies from sky coordinates
that other astronomers were examining. This is a loss of choice and a
handicap, yet SERENDIP has the virtue of forcing SETI astronomers
to examine star systems they might ignore. SERENDIP has been peri-
odically updated with improved equipment and software. SERENDIP
V, installed in 2009, examines 2.7 billion channels, whereas the first
SERENDIP covered only 100 channels.[43]

Billingham and Tarter's third category consisted of dedicated
searches, which had exclusive, permanent use of a telescope. One
prime example was the Ohio State University Big Ear search, which
would soon end. The problem was money. Ohio State University saw
greater utility in selling the twenty-acre site of the Big Ear telescope to
a golf club, which dismantled the radio telescope and replaced it with
a golf course in 1996.

The other prime example Billingham and Tarter listed was Paul
Horowitz's (b. 1942) SETI searches, which are still continuing. Horowitz has been able to raise money, perhaps because of his day job as physics professor at Harvard University and the fame of his popular text-
book, *The Art of Electronics*. Horowitz's SETI work has earned him the
scorn of Ernst Mayr, SETI's scourge, who accused Horowitz of wasting
his students' time. In SETI circles, however, he is a heavyweight.

Horowitz's SETI life began in 1981, during a sabbatical at Stanford. At
a NASA Ames meeting, Horowitz proposed looking for extraterrestrial

beacons. His plan was to focus on very narrow signals, which most likely would be artificial as opposed to the natural radio static of space. This strategy had the added bonus of simplifying the hardware. His proposal was immediately accepted. Horowitz designed a compact receiver that fit into three boxes and was dubbed "Suitcase SETI." Horowitz intended to carry it to available radio telescopes. In March 1982 he took his portable multichannel analyzer to Arecibo and scanned more than two hundred nearby sunlike stars.[44]

The Planetary Society

In his Arecibo venture, Horowitz had the financial backing of the new Planetary Society. It had been founded in 1980 by Carl Sagan, Bruce Murray, and Louis Friedman (b. 1941), also of the JPL. All three were space enthusiasts who feared that space exploration was boring Congress and the public.[45]

The Planetary Society was conceived as a citizens' advocacy group. Over the years, it has raised money to finance research overlooked by NASA, educated the public, and lobbied for space exploration. Although the aerospace industry also lobbies the federal government in behalf of space, the Planetary Society can point to its nonprofit status and claim purer motives. The Planetary Society has also supported SETI. Having helped Horowitz once, the Planetary Society helped him again when he sought to upgrade his SETI research.

After his allotted time at Arecibo, Horowitz had returned to Harvard, willing to travel again with "Suitcase SETI." Then, a stroke of luck. Less than an hour's drive from Harvard lay the obsolete Oak Ridge Observatory with its eighty-four-foot dish. With Planetary Society aid, Horowitz fixed it, built receivers, installed a new roof, and in March 1983 permanently housed Suitcase SETI at Oak Ridge. Expenses were minimal, about $20,000 a year for maintenance and electricity. Oak Ridge joined Big Ear as SETI's second dedicated radio telescope. In a contest to name it, the winning entry was "Project Sentinel."

Project Sentinel covered 128,000 channels, a stupendous number compared to the Ohio State search, but clearly not enough. There was always the possibility that ET was using an untapped channel. The solution was a new spectrum analyzer that could handle eight million channels. This required money.

In the United States, if a government grant is unavailable, another option is a foundation or a philanthropist. Carl Sagan linked Horowitz to Steven Spielberg, the movie mogul who had made science fiction films. His 1977 masterpiece, *Close Encounters of the Third Kind*, infused an extraterrestrial visit with the religiosity of a SETI true believer. If ufology has plagued SETI, the public conflating the two, now it helped the cause. Spielberg donated $100,000, and on September 30, 1985, a ceremony at the Oak Ridge Observatory christened META, or Mega Channel Extraterrestrial Array. Spielberg attended and quipped that he hoped "more [was] floating around up there than Jackie Gleason reruns." META was soon obsolete. The Planetary Society again helped Horowitz by contributing to Project BETA, which replaced META in 1995. One BETA improvement was immediate re-observation of a candidate signal. Unfortunately, BETA effectively ended in 1999 when a windstorm wrecked its antenna.[46]

The Planetary Society has also supported other SETI projects. In 1983 it contributed to an Australian SETI search. In 1990, Horowitz's META was replicated in Argentina and christened META II. As with the Australian search, astronomers in Argentina could reach portions of the southern sky not visible from North America. All these searches, however, were seen as preliminary to the big NASA program of 1992, which to the dismay of the SETI community lasted only a year.[47]

The SETI Institute

Congress's termination of NASA SETI in 1993 signaled a new era in American SETI. The big bucks of NASA were gone, enhancing both the private searches and the Planetary Society. The termination also gave a new mission to the SETI Institute of Mountain View, California, which unlike the Planetary Society has concentrated on alien life.

The SETI Institute was incorporated in 1984 by Jill Tarter and by Thomas Pierson (1950–2014), its CEO. Its original purpose was saving money for NASA SETI, which had to stretch its limited budget. NASA

SETI needed contractors, and they were expensive, coming from the West Coast universities who charged a huge overhead. The SETI Institute stepped in and searched for scientists and engineers at a discount, charging NASA less than the West Coast universities, even as little as one-fifth as much. According to Frank Drake, the president of the SETI Institute, by 1989 it was handling $2 million in contracts, besides running an education program in the schools.[48]

Pierson and Tarter had lobbied in Washington to save NASA SETI. When the roll call on Bryan's amendment began, Pierson was in his office at the SETI Institute, watching on C-SPAN. His initial optimism soon gave way and he said, "We're going to lose."[49] Literally, Pierson was correct. In a figurative sense, though, the SETI Institute had not lost; it was undergoing a transition—from SETI contractor to SETI player. Receiving NASA's unwanted search equipment as a "loan," the SETI Institute raised money and resurrected the target search, calling it Project Phoenix. A new era in the story of the SETI Institute and of SETI itself had begun.

9

After NASA

Revival, Distress, and Hope

Expelled from NASA, the SETI movement had to recoup, and it rose to the challenge. It developed new strategies. SETI@home pioneered in distributed computing by harnessing home computers to the search. OSETI, or optical SETI, represented a different direction, looking for light pulses instead of radio signals. Meanwhile, the SETI Institute evolved from its contractor origins, conducting its own searches as well as cheerleading for SETI. The Planetary Society remained supportive of SETI.

Survival aside, a case for pessimism in the post-Bryan years presents itself. The SETI movement seems to have crested in recent years. Outside the United States, SETI has little presence. In a June 2011 interview, Drake, the perennial SETI optimist, was downbeat, concerned over the lack of SETI careers for young people and over the lack of money.[1]

The big picture can also look gloomy, although in the short run the news has been good. Sagan, von Hoerner, and Shklovskii had feared that extraterrestrials had destroyed themselves and that humanity would soon follow either through nuclear war or overbreeding. So far, they have been wrong. The Cold War ended; the odds of the United States and the Soviet Union destroying each other and the planet have greatly diminished. Likewise, overbreeding, that other doomsday threat, no longer worries demographers. Although world population continues to increase, human reproductive rate is declining in most places. Sometime after 2050, world population is expected to peak and then fall.[2]

Why the Future Looks Ominous

Nevertheless, the L variable remains worrisome. Apocalyptic thinking may be hard-wired in the human psyche.[3] At one time, it was widely believed that supernatural forces would end our world. Today's secular age still believes in an Armageddon, but one whose agency is natural or human. New menaces have replaced overbreeding and the Cold War threat of atomic war. If Earth is typical—as SETI activists insist—these menaces might have already destroyed ET.

Science can go wrong. Sir Martin Rees (b. 1942), Britain's Astronomer Royal, and David Darling (b. 1953), British astronomer and author, have sounded the alarm.[4] Runaway chain reactions in a physics lab, self-replicating nanobots turning deadly, killer pathogens escaping from laboratories—these are among the doomsday scenarios. We can only speculate on extraterrestrial bad science.

Rees and Darling warn of Earth's natural forces. Super-volcanoes have caused mass extinctions. Some 74,000 years ago, on the island of Sumatra, the Toba event ejected rock, ash, and dust equivalent to the output of 10,000 Mount Saint Helenses. The vegetation die-off reduced the human race to fewer than 10,000, driving it to near extinction. Eventually, Earth's crust will produce another super-eruption. Are natural disasters the norm in the universe, and have humans so far been lucky?

Unfortunately, the universe is also capricious, dangerous like a battle zone where random shelling creates a lottery in which only the lucky, neither the virtuous nor the plucky, survive. Near-Earth Objects (NEOs) such as asteroids, meteors, and comets often hit Earth. Small NEOs are a nuisance; really big ones are something else. The huge piece of space junk that hit the Yucatán Peninsula 65 million years ago caused mass extinctions. Dinosaurs and many other species died off, and the Cretaceous period ended. Astronomers know that a huge NEO will again collide with Earth—in the distant future, we hope. The great SETI question remains unanswered. How many extraterrestrial civilizations have fallen victim to NEOs?

The effects of NEOs cannot be divorced from politics. Consider this scenario. In 1908, when a small asteroid or piece of a comet exploded over the Tunguska River area in Siberia and leveled more than two thousand square kilometers of forest, world leaders barely noticed. The impact area was remote, and the event remains a scientific curiosity. If the same impact had taken place fifty years later over Moscow, the surviving Soviet leaders might have hastily concluded that the Americans had attacked and quickly retaliated with their nuclear arsenal. The same Cold War fears might have affected American policy makers if Washington had been hit.

A gamma-ray burst from an exploding star also kills. Its high-energy radiation fries organic life within several hundred light-years. Solar flares are another danger. Although our sun is benign and its occasional flares have not been alarming, astronomers have noticed stars emitting bursts of intense radiation that destroy organic life. Both solar flares and gamma-ray bursts suggest a deadly universe. Earth has been lucky so far. What about ET?

How many extraterrestrial civilizations inhabit our galaxy? Carl Sagan believed a million; Frank Drake has lately been cautious, guessing at 10,000. Far fewer, said University of Washington professors Peter Ward (b. 1949) and Donald Brownlee (b. 1943) in their 2000 book *Rare Earth: Why Complex Life Is Uncommon in the Universe*. Ward and Brownlee argue that Earth is unique, the beneficiary of many lucky breaks. They call Earth a "Goldilocks" planet because much of it is just right. Earth is the right distance from the sun; if not, it would be too hot or too cold. The moon has the right size; if not, Earth would suffer an irregular orbit and huge swings between seasons. Even the solar system is just right. The enormous gravity of Jupiter, whose mass is three hundred times Earth's, either deflects or sucks in space junk that otherwise might hit Earth. Ward and Brownlee pile up the lucky circumstances and conclude that complex organisms are unlikely beyond Earth. They do concede that simple life such as bacteria might be common.[5]

The SETI Institute

The SETI Institute must either discount the case against ET or close shop. The cancellation of NASA SETI gave an excuse to surrender, but instead the SETI Institute rebuilt from the ruins. The all-sky search at the JPL was out of reach; it used equipment also developed for NASA's Deep Space Network. SETI was luckier with the targeted search at Ames. NASA did not need the equipment and lent it on a long-term basis to the SETI Institute. Inspired by the mythological bird that rose from its ashes, Barney Oliver recommended the resurrected target search be named Project Phoenix.

As always, money was necessary, and Oliver came to Project Phoenix's rescue. A Silicon Valley kingpin, he knew the right people. Bill Hewlett and Dave Packard, his former employers, each donated $1 million, as did Gordon Moore of Intel Corporation, Microsoft co-founder Paul Allen, and Mitchell Kapor, founder of Lotus. Not all the solicited contributed. Ted Turner replied that he and his wife, Jane Fonda, were concentrating on the environment. Rupert Murdoch was uninterested. In all, including himself and many other contributors, Oliver raised $7.5 million.[6] His final service to the SETI Institute came after he died in 1995. He left it $20 million.

Project Phoenix started in February 1995. It ended nine years later, after eleven thousand hours of observations at the sixty-four-meter Parkes telescope in Australia, the new Green Bank Telescope in West Virginia, and Arecibo. Under Jill Tarter's direction, the project concentrated on eight hundred sun-like stars within 250 light-years. For each star, Phoenix examined some 2 billion channels in the microwave band between 1.2 and 3.0 gigahertz. Phoenix was listening for narrowband transmissions, hoping that extraterrestrials were transmitting beacons in our direction.

Project Phoenix remains the most extensive SETI search ever. It duplicated Drake's two hundred hours on Project Ozma in one tenth of a second. It failed, but each star system received only one day of intense scrutiny. Bad luck was always possible. ET's planet may have been offline when Project Phoenix was tuning in.[7]

The SETI League

All-sky searches avoid the offline problem. With the equipment of the defunct NASA all-sky search unavailable to SETI researchers, imagination was needed. It came from New Jersey industrialist Richard Factor. In 1994 he established the nonprofit SETI League, whose Project Argus ambitiously invites public participation in the search. Unlike Project Phoenix, which was establishment astronomy, the SETI League recruits amateur astronomers willing to develop technology and software to search the skies with backyard radio telescopes, typically three- to five-meter dishes—all at their own expense. The SETI League coordinates their efforts.

The SETI League, under its executive director, H. Paul Shuch (b. 1946), had envisioned five thousand radio telescopes for real-time coverage of the entire sky, hence the name "Argus," the mythical Greek beast with a hundred eyes that could see everywhere. The first five stations went online on Earth Day, 1996, but growth has been slow. By 2003 volunteers had erected only 123 telescopes in 23 countries, and by 2014 Project Argus had practically stalled, with only 147 participants in 27 countries.[8] Unlike Argus, the eponymous project will not see everywhere. The pool of electronics buffs willing to spend several thousands of dollars to assemble the equipment seems limited.

As of 2011 the SETI League had only 1,548 members in sixty-two countries, and its finances are depressing. Revenues hit a high of nearly $247,000 in 1999 and have steadily shrunk to $16,000 in 2014.[9] The current economy is not the villain. The league's financial declension began after 1999 when the economy was still strong. The root cause of the malaise may be that SETI needs an extreme willingness to delay gratification. Unlike dieting, whose payoff is readily apparent at the scale, contact has no timeline. Contact may not happen in one's lifetime. Only a special kind of person can handle the continuous failure.

SETI@home

SETI@home is another imaginative scheme. Requiring less time and money from individuals, it has been more successful, a good example of supply meeting demand. The demand factor is SETI's enormous need for computing power. SETI must identify the millions of signals hitting Earth and discard those of natural origin or from Earth's satellites. To complicate matters, if ET sends a signal, its parameters will be a mystery. The frequency, the bandwidth, and the duration are unknown, as are other parameters, such as whether the signal is being beamed directly to Earth or in all directions. The signal's distortion due to Earth's rotation also complicates, and the rotation of ET's world is a total mystery. Supercomputers can investigate these parameters, but SETI research does not enjoy the cachet of medical research or cryptography. The supercomputers of government and industry are unavailable.[10]

The solution has been a new supply source: distributed computing. The idea goes back to the early 1980s when scientists at XEROX PARC linked one hundred Alto computers via the first Ethernet. Like much of the innovation at the fabled XEROX PARC, its fruition took place elsewhere. In 1996, computer scientists at UC Berkeley began developing software to connect the unused power of home and office computers and create a virtual supercomputer.[11] As always, funding was necessary. Silicon Valley was an obvious source, but SETI@home, as the system came to be called, did not fit the standard business model, for it was developing free software. In 1998, project director David Anderson and project scientist Dan Werthimer received $50,000 from the Planetary Society. Another $50,000 came from Paramount Pictures, which was publicizing its science-fiction film *Star Trek Insurrection*. Not for the first time, the public's fascination with space adventure has benefited SETI.[12]

Werthimer and his colleagues envisioned SETI@home as a short-term project of 200,000 to 300,000 participants. They did not anticipate the huge appeal of a scientific quest that demanded little. Participants downloaded the software through the Internet and then sat back while their computers processed Arecibo data from the UC Berkeley server and automatically returned them. After its launching on May 17, 1999, three million SETI buffs in 223 countries signed up within three months. When SETI@home reached its original termination date of May 2001, it was extended. Within three years, SETI@home had processed 500 million packets of data. By 2003, SETI@home boasted 4.7 million volunteers.[13]

Besides its ease, SETI@home offered people a chance to save the planet. Over half of the users surveyed believe that contact will benefit humanity. Not to be discounted was the prospect of fame. SETI@home promised to share honors with the person whose computer first detected an alien signal. Like a lottery in which additional tickets increase the odds of winning, 40 percent of participants were using two or more computers. Clubs were formed to pool resources, creating competition among individuals and clubs for the most scanning. A minority of less than 1 percent, extremely eager to win, cheated. When SETI@home noticed tampering with returned data and with the program's settings, it had to waste time checking.[14]

SETI@home has grown sophisticated over the years. In 2004 it introduced the BOINC (Berkeley Open Infrastructure for Network Computing) software, which allows other scientists to use distributed computing in fields ranging from climate study to protein folding. In 2009, eighty projects were using this software. Although SETI@home has competition, it remains the most popular of the donated computer time projects. As of April 2010, more than 8 million had volunteered their computers.[15] SETI@home has outgrown Arecibo; in 2011 it was also using the Robert C. Byrd Green Bank Radio Telescope in West Virginia.[16] As of June 2013 there were 145,000 active computers from 233 countries in the system.

SETI@home does have challenges. Like SERENDIP, SETI@home suffers from being a piggyback operation. Its data come from portions of the sky that other astronomers at Arecibo are studying, not necessarily portions the SETI people prefer. Furthermore, SETI@home does not check in real time; data are examined long after their reception, precluding an immediate follow-up. Besides, SETI@home looks at a narrow frequency band of 100 MHz centered at the hydrogen line of 1,420 MHz. ET may be broadcasting outside of this band.[17] SETI@home's greatest challenge is money. Werthimer was hoping Australia's Parkes radio telescope would cover the Southern Hemisphere, but this was contingent on funding. As of 2014, the Parkes telescope was unavailable.

The SETI Institute and the Allen Telescope Array

It was obvious back in the 1990s that SETI needed a dedicated radio telescope. Project Phoenix had its virtues, but being always online was not one. It had to wait its turn at radio telescopes around the world. In its last six years at Arecibo, it scanned only 5 percent of the time.

With NASA out of the picture, the dedicated telescope had to be privately built. In January 1999 the SETI Institute and the Radio Astronomy Laboratory at UC Berkeley announced an imaginative partnership. In return for the SETI Institute's funding and constructing an array of 350 six-meter antennas, named the One Hectare Telescope, or 1HT, the Radio Astronomy Laboratory promised to design and operate it. The 1HT's locus was UC Berkeley's Hat Creek Observatory in the secluded Hat Creek Valley, about a five-hour drive north of San Francisco.

The array was cheaper than a large single-dish telescope. When completed, the 1HT would rank among the world's largest, able to catch transmissions from as far away as a thousand light-years. By agreement, astronomers at UC Berkeley received 20 percent of the observing time. The good news for the SETI Institute was that it could use the array while Berkeley was doing conventional radio astronomy. Compared to Project Phoenix, the array's range of frequencies was advantageous. Ranging from 500 to 11,500 MHz, they were six times greater than Project Phoenix's.[18]

Without funding, the 1HT was fated to remain on the drawing board. Luckily, Paul Allen (b. 1953), a contributor to Project Phoenix, was a science-fiction buff, and he could not resist the siren call of ET. He contributed $25 million. A grateful SETI Institute renamed the array, and it is now called the Allen Telescope Array, or ATA.

Allen's generosity was insufficient. The SETI Institute fell short of its fund-raising goal for the array. Naming it after Allen may have been premature, and a mistake, for another big contributor could not be memorialized. The ending of the 1990s dot-com boom did not help. Silicon Valley had less money for worthy causes.

Moreover, the nature of philanthropy was changing. According to the *New York Times*, philanthropy has become "outcome oriented" and seeks "clearly defined goals." Donors and grantees monitor progress "in order to make appropriate course corrections."[19] This is not the SETI way, in which astronomers rule and fast results are unlikely. A philanthropist with long-lived ancestors will sooner die old than see contact. Not accidentally, Bill Gates and Warren Buffet are using their fortunes to eradicate diseases, a campaign whose results are both immediate and quantifiable on a spreadsheet.

Meanwhile, the ATA was suffering cost overruns and delays. The designers at UC Berkeley kept on suggesting revisions, and the SETI Institute, eager for the perfect telescope, acquiesced. The cost kept on ballooning. Unable to finance the full array, the SETI Institute finally decided to go ahead and use the $50 million it had raised to construct forty-two dishes. In 2007 the Allen Telescope Array, in its diminished state, was dedicated. Allen pushed a silver button, and the forty-two antennas went online.[20]

Incompletion was only the start of the Allen Telescope Array's woes. In April 2011 the ATA temporarily closed down when the Hat Creek Observatory, which housed the array, ran short of money. The setback began with the National Science Foundation (NSF), whose grants partially funded the observatory. In 2008 the NSF cut this funding by 90 percent, claiming that the incomplete ATA was incapable of doing the envisioned radio astronomy. The final blow came in 2011, when the budget-challenged State of California drastically reduced its allocation to UC Berkeley's Radio Astronomy Laboratory, which managed the observatory. In April the Allen Telescope Array stopped taking data.[21]

The ATA faced permanent closure if the SETI Institute could not find another revenue stream. To restart the ATA for a year, the SETI Institute needed $1.5 million for operations and $1.0 million for the scientific staff. By December, an appeal to the public raised over $230,000 and the ATA reopened. The twenty-five hundred donors included Jodie Foster, star of the movie *Contact*, which was based on Carl Sagan's novel.[22]

Donations from the public made good publicity, but they were really Plan B. Plan A was the U.S. Air Force. When the NSF reduced its grant, the SETI Institute had seen the proverbial handwriting on the wall. In 2009 it was negotiating with the Air Force Space Command. The

Air Force's Space Surveillance Network needed additional telescopes to track space debris that could damage defense satellites. Satisfied with the ATA, the Air Force finalized the deal in April 2012. In return for tracking satellites some of the time, the ATA receives financial security. The new revenue stream allowed replacing the cash-strapped Radio Astronomy Laboratory. SRI International, a nonprofit independent research center, assumed management of the Hat Creek Observatory.[23]

As of 2010, big donors had apparently abandoned the SETI Institute. In the *SETI Explorer*, John Gertz, chairman of the board, wrote that the SETI Institute could not "rely on being the playground of a few hyper-wealthy donors because they . . . can shift focus or their financial situation can change." Was Gertz referring to Paul Allen? According to Gertz, the institute needs "a very broad base of individuals" who come "from a wide range of nationalities, cultures, and demographics." In other words, the SETI Institute must rely on many small contributions.

Maybe Gertz spoke too soon. Two years later, Qualcomm cofounder Franklin Antonio donated $3.5 million to upgrade the ATA. It is significant that the money was not used to add dishes to the telescope.[24] In any case, when Jill Tarter retired from the SETI Institute in May 2012 she said she would spend her golden years raising money for SETI.[25]

Unfriendly Politicians

The chances of NASA's resuming SETI research are slim. In 2001 the House Subcommittee on Space and Aeronautics held a hearing on SETI. Testifying in behalf of SETI were Chris Chyba of the SETI Institute; Jack Farmer, an astrobiologist at Arizona State University; NASA's Ed Weiler; and Neil deGrasse Tyson (b. 1958), director of New York's Hayden Planetarium and Carl Sagan's successor as the public face of science. Representative Sheila Jackson Lee, a Democrat from Texas, was not impressed. She asked about practical spin-offs in medical research. When Chyba spoke of educational effects and of a possible spin-off in medical imaging, Jackson Lee replied that "practical applicability" was necessary to move forward. As with Senator Bryan, butter prevailed over stars.[26]

The SETI community is not giving up. On December 4, 2013, Steven Dick (b. 1949), a retired NASA historian, gave an overview of astrobiology and exoplanets to the Committee on Science, Space, and Technology of the U.S. House of Representatives. Adding a few remarks on SETI, he called for a reinstatement of federal funding that would "repair the artificial programmatic divorce between the search for microbial and intelligent life." As of early 2016, that funding has not been restored.[27]

SETI's woes reflect the greater woes of planetary science. If Bernard Oliver were alive, what would he say? Oliver may have been the most enthusiastic SETI booster of his day, convinced that contact outranked the Incarnation. In 1975, decrying the government's disinterest in SETI, he told the U.S. House Subcommittee on Science and Technology that the future, in looking back at the twentieth century, would not understand "the obsession with defense and social programs."[28] The social spending he opposed from his one-percenter perch is higher than ever. Astronomy suffers. In 2006 the NSF recommended slashing Arecibo's budget.[29] Astronomers were concerned enough to launch a campaign to save the observatory. Arecibo survived but enjoys less federal funding and must now compete for grants.[30] The Arecibo Observatory, so important to Project Phoenix, SERENDIP, and SETI@home, as well as conventional astronomy, faces an uncertain future.

Arecibo is no isolated case. In February 2012 the Obama administration's proposed budget for fiscal year 2013 reduced NASA's planetary science program by 20 percent, forcing NASA to cancel partnerships with the European Space Agency. The Planetary Society warned of the effects; the flagship programs, famous for exploring the solar system, sending rovers to Mars, and spacecraft to the outer planets, would end. After space advocates lobbied, the administration partially relented, announcing in March 2013 that it would restore most of the cuts. Space exploration has remained on the defensive.[31] On March 24, 2014, the Planetary Society again sounded the alarm. It informed members by e-mail that the White House's proposed budget for the following year cut hundreds of millions of dollars from NASA, "including for the third year in a row—Planetary Science." Again, space advocates lobbied

Congress, which restored the cuts and increased NASA's budget by 2.5 percent. For SETI, tight budgets for NASA mean that Senator Bryan's termination of NASA SETI will endure.[32]

The Obama White House might have been reflecting public attitudes. In the 1960s, Cold War fears fueled the higher interest in space, which abated when the Soviet Union collapsed. If membership in the Planetary Society is indicative, the public is losing interest in space. In 2011, revenues from dues were 25 percent less than in 2001; in 2012 they were off by 34 percent—still better than 2009 and 2010, when dues were off by 38 percent. In the 2013 fiscal year, total revenue was above 2012's but still below the years from 2009 to 2011.[33]

The Uncertain Outlook

As for contact, what Shklovskii called the "adolescent optimism" of the early years is gone. No longer is contact seen as the simple matter of identifying ET's frequency. Spectrum analyzers now scan millions of frequencies every second and find nothing. Failure has made the SETI community cautious. The SETI Institute's 2002 blueprint for the future predicted the search taking "years, perhaps decades and possibly centuries."[34] Tarter admits that contact might not occur in her lifetime. Werthimer believes it will take fifty to one hundred years. In a 2010 interview, Drake was guarded, unlike the past when he predicted contact coming sooner than later. Pressed for a date, he suggested twenty to thirty years, carefully adding that he was guessing.[35]

Seth Shostak (b. 1943), the unofficial spokesman of the SETI Institute, admits the outlook is uncertain. In 2011 he remarked that the White House kitchen staff is larger than the total number of SETI employees worldwide, who are mostly in the United States.[36] Only three other countries have continuous SETI programs. In 1990, with Planetary Society backing, Guillermo Lemarchand (b. 1960) started a SETI program at Argentina's Instituto Argentino de Radioastronomía. In Australia, the Parkes Observatory has been running a piggyback SERENDIP survey since 1998. Neither project is cutting-edge. The Argentine search uses two thirty-meter dishes with low sensitivity, and the Australian antenna has only 4 percent of the collecting area of Arecibo. Yet both Argentine and Australian SETI are valuable; their southern locations allow scanning portions of the Milky Way that telescopes in the Northern Hemisphere telescopes cannot reach. Europe, from which much would be expected, has only one continuous SETI research project. SETI Italia piggybacks a version of SERENDIP IV onto a thirty-two-meter dish in Medicina near Bologna.

Why do non-American astronomers ignore SETI? They have the resources and knowledge. Shostak wonders whether the answer is the United States' "frontier mentality, the willingness to gamble on a long shot with a big payoff."[37] Other explanations are possible. Maybe the size and diversity of the United States facilitate nonconformity and permit a movement that many consider quixotic. Perhaps the explanation could be the "great man" factor. American SETI is almost inconceivable without Drake and Morrison, or with Shklovskii and Kardashev in the Soviet Union.

The case against SETI dates to Project Ozma and continues to be updated. The Drake Equation has long been criticized for generalizing from a single example. Princeton researchers David Spiegel and Edwin Turner restated this criticism in a January 2012 paper. Using Bayesian analysis, they argue that optimistic assumptions instead of hard data underlie the belief in extraterrestrial intelligence.[38]

British physicist Stephen Webb echoed Enrico Fermi and Michael Hart by again asking, "Where are they?" Radio signals are only one sign of intelligent life, he noted. Advanced extraterrestrials should leave other traces of their presence. Yet, Webb asks, where are the infrared emissions from Dyson spheres, the artificial lines in stellar spectra, signs of vehicles that burn antimatter, and for good measure the Bracewell probes? None of these has been detected.[39]

If extraterrestrials are not intentionally sending us radio signals, SETI has hoped nonetheless to detect leakage, the alien equivalent of military radars and TV signals. This will not happen, though, if Earth is typical. New technologies are silencing Earth. Cable TV emits no signals. Digital signals need less power and require decoding. The extraterrestrials who will receive 1950s television will not receive the recent

Super Bowl. If advanced extraterrestrials are leakage free, Paul Davies suggests, SETI's hopes rest on extraterrestrials who want to exchange information, whose societies prize knowledge for power and wealth. This reasoning assumes a Galactic Club of curious extraterrestrial civilizations.[40]

Steven Dick gives an ingenious reason for the lack of contact. It is the extreme sophistication of many extraterrestrials. He faults SETI "practitioners" for ignoring cultural evolution and assuming that extraterrestrial civilizations have a developed form of Earth's current technology. Using human progress as his model, Dick projects the growth of artificial intelligence beyond a thousand years and sees humanity transitioning to a postbiological existence. If humanity is typical of galactic life, advanced extraterrestrials have already reached this stage. Extraterrestrial postbiologicals are both machines and super-intelligent, and in very old civilizations have "characteristics approaching those we ascribe to deities: omniscience [and] omnipotence." They may not want to communicate with us biologicals, finding us benighted, backward, and boring. The vast differences make mutual comprehension impossible. Dick concludes that SETI will succeed only with civilizations less than a thousand years advanced.[41]

Active SETI

If there are extraterrestrials within Earth's technological range, it is possible that they are self-satisfied and complacent, with no desire to transmit. They may need a nudge, a message from Earth to begin a dialogue. This is the reasoning behind Active SETI, also known as METI (Messaging to Extraterrestrial Intelligence). Russian radio astronomer Alexander Zaitsev (b. 1945) has led the way. Using the Yevpatoria Radio Telescope in Ukraine, Zaitsev has beamed four messages to thirteen nearby star systems. In 1999 he sent *Cosmic Call 1*, which among other things included an elementary encyclopedia. Two years later, his *Teen Age Message* transmitted Russian and English greetings as well as music from a theremin (an electronic musical instrument). *Cosmic Call 2*, in 2003, contained contributions from Russians, Americans, and

Canadians. *A Message from Earth* in 2008 beamed 501 missives selected through an Internet competition.[42] Although Zaitsev refers to interstellar dialogue, his messages speak to an Earthly audience.[43]

Other Active SETI messages outdo Zaitsev in showmanship. On February 4, 2008, NASA beamed the Beatles song *Across the Universe* to Polaris (the North Star). NASA was marking both its fiftieth anniversary and the fortieth anniversary of the song's recording. Since Polaris is 431 light-years distant, the reply from one of its planets, if sent, will arrive in 862 years. NASA was either seeking publicity or creating an electronic memorial of sorts. On June 12 of the same year, the University of Leicester sent a Doritos commercial, which the British public selected, to the Ursa Major constellation.[44] A stunt to publicize junk food posing as interstellar messaging! On August 15, 2012, the National Geographic Channel promoted its new series *Chasing UFOs* when it transmitted in the direction of 1977's "Wow!" signal. It sent more than ten thousand Twitter messages as well as videos from celebrities such as 2011's Miss Universe and comedian Stephen Colbert, who assured aliens that human meat is gamey and they should instead eat at the Crab Nebula.[45] Will ET appreciate Colbert's humor?

Active SETI has sparked a lively debate regarding whether Earth should announce its presence to the galaxy. Zaitsev defends it: "Perhaps, after 50 years of listening to nothing but cosmic static, it is time to recognize that the time has come for humankind to take the lead in helping to end the Great Silence. Could it be that the future of SETI lies not in receiving, but rather in transmitting?"[46] Douglas Vakoch (b. 1961) of the SETI Institute adds that Active SETI places the burden of decoding on an advanced civilization, which presumably will have more success than Earth's cryptologists.[47]

David Brin of "The Mystery of Great Silence" is skeptical. He accuses the SETI community of suffering "frustration at the lack of contact" and of losing patience. "The decades of silence" have taken "a toll." Brin offers a reason for the Great Silence: something destroys civilizations who call attention to themselves. Why yell in the jungle and alert predators, he asks. For good measure, Brin dismisses Zaitsev as a remnant of the Soviet era, which defined all advanced societies as socialist and altruistic.[48] Another view is that the Soviet Union is defunct, but the empire manages to strike back. The Russian science phalanx is marching to its own tempo, mirroring Vladimir Putin's rejection of the West.

Stephen Hawking echoes Brin's caution. Pointing to human history and the aggression inflicted on technologically inferior civilizations, he suggests that Earth should "lay low." Hawking has furthermore warned of extraterrestrial predators roaming the galaxy in large ships searching for lush planets to plunder.[49] This scenario is suspiciously similar to the plot of the 1996 science-fiction movie *Independence Day*.

Sagan and Drake did not take chances when they transmitted the first Active SETI message in 1974; they chose a destination 25,000 light-years distant. By contrast, the Active SETI transmissions of the Zaitsev era lack this caution. Zaitsev beamed to thirteen constellations. The farthest was Delphinus, which will receive his *Teen Age Message* in 2070. The nearest was Libra, which should receive his *Message from Earth* in 2029. The Doritos commercial was transmitted to a solar system forty-two light-years near. If these messages cause extraterrestrials to take umbrage, many of today's people will still be alive.

The Active SETI people are unfazed. They reply that TV and radar transmissions have already blown Earth's cover. Shostak states that extraterrestrials advanced enough to hurt us are also advanced enough to detect Earth's transmissions, yet nothing has happened. Zaitsev notes that the powerful radar signals emitted into space far outnumber the few messages deliberately sent to extraterrestrials.[50] This may be so, but Zaitsev conveniently ignores a possibility. If radar signals have alerted extraterrestrials, why do they remain silent?

Opponents of Active SETI offer a reason for this silence. Earth's cover has not been blown, they say. The reason is simple. Military and research radars have not been directed to a specific location. The vastness of the universe makes it likely that these signals have missed planets. On the other hand, Active SETI transmissions are targeted; they will hit planets, which may be inhabited. As for weaker TV signals, extraterrestrials catch them only if they have built huge receivers far surpassing anything on Earth.[51]

Does this debate really matter? It does if aliens have their own SETI programs. But if they are not listening when Active SETI is transmitting or are tuned on a different frequency, they will not catch the signal.

If they should catch it and are confused, they might look for a second signal. It will not come, since Active SETI signals are not continuous; they have lasted less than an Earth day. The aliens might dismiss the lone signal as a mysterious anomaly—their version of the "Wow!" signal.

Active SETI succeeds only if the extraterrestrials receive the signal and recognize its unnatural origin. Aliens with a culture gap might never decipher the signal and stay forever ignorant of the Doritos cuisine. Whether they decipher or not, the aliens can respond by sending a beacon in our direction or they might not. They might lack curiosity or they might be fearful.

A violent response is possible. The worst fears of Brin and Hawking may be realized. No one knows. Here is another example of humanity gambling with the future. A comparison can be made to genetically modified crops whose safety and environmental impact have been questioned. Boasting a superior yield and pest resistance, these crops are grown in many parts of the world. We hope that these "franken-foods" are harmless. Active SETI is just as problematic: a roll of the dice that extraterrestrials are benign. All we can do is hope, as with genetically modified crops, that the alarmists are wrong. There is a greater philosophical question. As anthropologist Kathryn Denning puts it: "What is the right way to balance the desires of some people against the concerns of others, and who is entitled to make decisions about the future of the world which we all share?"[52]

The Drake Equation Again

There is no doubt that the SETI community has endured a litany of bad news. Yet SETI has reason to hope. The best news is that the Drake Equation, which has always been problematic, a triumph of optimism, is no worse.[53]

The bad news concerns L, the key variable for the SETI community. The lifetime of communicating civilizations is more uncertain than ever. Whereas the Green Bank Ten fretted over overpopulation and war, we realize today that natural disasters can also destroy extraterrestrials. To believe that alien civilizations survive for millions of years, once the politics and demographics are fixed, appears somewhat naive.

As for the f_c variable, the fraction of planets capable of interstellar communication, some updating is needed. The SETI people once presumed that advanced extraterrestrials were using radio, even though it is a twentieth-century technology: an optimism, to say the least, that may explain the failure to contact ET. This failure, though, has made SETI realize it cannot rely solely on radio signals.

Frank Drake has said that SETI research began with radio simply because radio technology was well known. Not surprisingly, since then the SETI community has embraced the technology of the twenty-first century. Optical searches now complement radio. Yet, using light for interstellar communication is not a new idea. In the nineteenth century, as already noted, it was suggested that Earthlings light huge fires to signal Martians.[54]

Modern optical SETI, or OSETI, was born in 1961. Charles Townes, inventor of the laser, published a paper (R. N. Schwartz, coauthor) which pointed out that extraterrestrials might be using laser beams to communicate. Bernard Oliver was not impressed. Viewing extraterrestrial technology through the prism of the 1970s, Oliver believed a continuous, energy-sucking laser signal was necessary to outshine a central star, causing extraterrestrials to balk at the huge power needs. Oliver's prestige resulted in OSETI being ignored in the next two decades.[55]

By the 1990s, OSETI was being reconsidered. Advances in laser technology made a continuous laser wave unnecessary; a short burst of light could do the job. Lasting a fraction of a second, a laser burst could outshine the sun by a factor of more than ten thousand. Best of all, the power consumption was tolerable. If Earth can do it, so can ET.[56]

An optical signal has advantages over radio. Its higher frequency carries more information. It is immune to the background noise, human and celestial, that bedevils radio. Sent in a tightly focused beam, a laser beam is less likely to dissipate than radio waves.[57] Even so, nothing is perfect. Unlike radio, laser cannot penetrate clouds of interstellar dust and other materials. Optical messaging also has an eerie aspect. It must be directly targeted. Extraterrestrials transmitting laser bursts to Earth know we exist. Drake says: "That raises a whole new issue, which is whether there is altruism in the universe."[58]

OSETI is the promising search technology of the twenty-first century. Both Dan Werthimer and Paul Horowitz have jumped on the OSETI bandwagon. Werthimer's project is called SEVENDIP, for Search for Extraterrestrial Visible Emissions from Nearby Developed Intelligent Populations. Launched in 1997, SEVENDIP uses a thirty-inch automated telescope in Lafayette, California, to scan the sky for a nanosecond pulse.

In 1998, Horowitz began a piggyback targeted search at Harvard. It ended in 2004, when funding ceased. OSETI needed something better than a piggyback; it needed a dedicated telescope. With the aid of the Planetary Society, Horowitz built one, which has the added advantage of being all-sky, thus increasing chances of catching a light burst. As of 2013, the telescope is analyzing trillions of bytes per second.[59]

These are not the only OSETI programs, but they suffice to show that the SETI community is open to other avenues. It realizes that ET may be using a different portion of the electromagnetic spectrum. As knowledge of the entire spectrum increases, searches of the infrared and ultraviolet bands may also lie within the range of human technology.

Although OSETI makes f_c seem stronger today, Gerrit Verschuur's take on the technological variable cannot be ignored. Verschuur notes human technology's dependence on energy. Without fossil fuels, civilization, let alone technology, would have been impossible on Earth. Moreover, according to Verschuur, lodestone (a naturally occurring magnet) and amber (which generates static electricity) led to human awareness of magnetism and electricity. If these substance are not readily present on other worlds, the odds against alien technology (or, at least, one remotely resembling ours) increase.[60]

The n_e variable, the number of a star's planets with conditions conducive to life, seems encouraging. Alien life was once seen needing an environment favorable to earthly life, as it was traditionally known. This anthropocentric delusion ended in the 1980s with the discovery of extremophiles—organisms that thrive in extremes of temperatures, radioactivity, gravity, or acidity once believed toxic to all life. Some examples are worms living by volcanic vents near the ocean floor, bacteria thriving in nuclear reactors, and slimes two miles under the surface consuming rock instead of organic matter.[61]

Extremophiles give a wider range of life-friendly worlds to SETI research. A strange life may lurk beneath the Martian surface. Also intriguing are the large, icy moons of Jupiter and Saturn. Europa, Titan, Ganymede, Callisto, and Enceladus may hold surprises beneath their icy crusts. Far from the sun, these moons are dead to traditional life, but they may harbor weird life. Volcanic vents beneath their icy surfaces could provide energy to unusual life-forms. One wonders what lies beyond the solar system.[62]

The presence of extremophiles on Earth suggests that life is possible elsewhere, not that it does in fact exist. The value of the Drake Equation's f_l variable, the fraction of life-bearing planets in a star system, remains a mystery. Nevertheless, this variable shows more promise today.

Other worlds may harbor life, but is this life intelligent? Although the SETI community has traditionally regarded L as the key variable in the Drake Equation, f_i, the fraction of planets that have intelligent beings, may be the one that really counts, and it remains a mystery. Because extremophiles raise the chances of primitive life beyond Earth, the possibility of this weird life growing complex and perhaps intelligent must be considered. On the other hand, the biologists' case against a widespread intelligence cannot be discounted.

The variable in the Drake Equation exciting the SETI community these days is f_p, the fraction of the galaxy's stars that form planets. For most of SETI's history, the existence of extrasolar planets was assumed. Although not detected, they simply had to be there, undetected, or else SETI was pointless.

In 1995, extrasolar planets finally transitioned from an undetected necessity to fact. Swiss astronomers Michel Mayor and Didier Qeloz discovered 51 Pegasi, the first exoplanet around a sunlike star, which, unlike earlier "discoveries," passed peer scrutiny. Pioneers are important because they encourage others. Drake's Project Ozma was primitive, but it ushered in the SETI age—likewise after Mayor and Qeloz. As of September 2015, NASA has identified more than 4,600 possible exoplanets.[63]

Search methods for exoplanets initially found very large planets, because exoplanets are not literally seen. They are lost in the brightness of their parent stars. A comparison is trying to see a gnat next to a 250-watt light bulb. As a result, exoplanets are detected indirectly. In the radial-velocity method, astronomers look for a star's "wobble" when a planet's gravity affects it; the larger the planet, the more the parent star wobbles. Size does matter. Size also matters in the transit method. A star dims when a planet transits in front of it; the dip in brightness depends on the size of both the star and its planet.[64]

Because size matters, gas giants, larger than Jupiter, were the first exoplanets discovered. Found very near their stars, they were dubbed "hot Jupiters."[65] Unfortunately for SETI research, Jupiter-type planets are not conducive to life remotely resembling ours. SETI needs Earth-size, rocky planets with liquid water and in regular orbit around a sunlike star. Luckily for the SETI community, astronomers developed detection methods of greater sophistication, allowing the discovery of smaller planets. By 2004, so-called super-earths, ten times the size of our planet, were being found.

To locate exoplanets, the European Space Agency launched its Corot mission in 2006, and NASA followed with the Kepler spacecraft in 2009. By July 2015 the Kepler Space Telescope, which uses the transit method, has confirmed 1,039 exoplanets in the Cygnus and Lyra constellations. Twelve of the exoplanets not only lie within the habitable zone but are also Earth-sized, that is, are less than twice the size of Earth.[66]

Although Kepler stares at only 150,000 of the galaxy's 300 billion or so stars, extrapolations are possible. According to a statistical projection of Kepler data published in 2013, the galaxy contains 40 billion Earth-like planets. One caveat is in order. Although roughly the size of Earth and within the habitable zone, these planets may not be life-friendly.[67] Nonetheless, the f_p variable is today far more SETI-friendly than it was for the Green Bank Ten.

Kepler's Twilight

For a while, funding issues aside, SETI was on a roll. The days when radio telescopes scanned sunlike stars, in full ignorance whether these stars hosted planets, seemed over. Thanks to Kepler, SETI researchers could increase the odds of success by looking only at Earth-size planets in habitable zones. Even better would be avoiding lifeless planets altogether by first spotting biosignatures, chemical signs of life such as oxygen. NASA had planned precisely this with its Terrestrial Planet Finder (TPF) mission, but since 2011 the mission has been on hold. NASA is diverting scarce funds to the James Webb Space Telescope, successor to the Hubble telescope, and scheduled for a 2018 launch. Many astronomers had resented the TPF mission, preferring non-planetary projects such as studying the Big Bang and the evolution of galaxies. One scientist heavily committed to the TPF sees the project being taken off the shelf only in the far future—decades, if ever.[68]

As if scientific infighting and Washington budget cutting are not enough, a setback beyond human agency struck the Kepler Space Telescope. Originally given a three-year duration, Kepler's mission was extended to 2016, but the telescope broke down. Kepler used four reaction wheels to aim at a target. One wheel failed in 2012, but the telescope could still observe. In May 2013 a second reaction wheel failed. NASA tried to repair the wheel, but in August it admitted defeat. William Borucki, Kepler's director, was hopeful that the remaining functional parts of the telescope could do other research. He was right. In May 2014, NASA approved a limited, and somewhat new, mission for Kepler, called K2. Using the sun's rays to help in steering, Kepler repositions every seventy-five days and observes varied phenomena such as red dwarfs, supernovae, and galaxy clusters. Although in its K2 mode Kepler remains useful to astronomy, Kepler is less useful to SETI. As a consolation of sorts, the accumulated data from Kepler's original mission will occupy researchers for three or four years.[69]

So frustrating for the SETI community! So much technology today would be almost science fiction for the SETI people of the 1960s. Yet the Kepler spacecraft has let SETI down. NASA has dropped the Terrestrial

Planet Finder mission, refusing a closer look. There may be a bright note, though. In September 2013, NASA announced it had tweaked the Spitzer Space Telescope. Launched in 2003, Spitzer focused on infrared emissions to study stars and comets. Spitzer's new mission is using its infrared technology to assess the temperature and climate of exoplanets.[70] Whether Spitzer will help SETI remains to be seen.

The SETI Institute Adjusts

The SETI Institute and the Planetary Society—indeed, the entire SETI community—must be nimble, often forced to cope with events beyond their control. The SETI Institute has evolved from its origins as a contractor for NASA. It reacted to the Bryan debacle by initiating its own SETI searches. Although searching remains the primary mission, the SETI Institute has branched into astrobiology and planetary research. In 2014 the SETI Institute's Carl Sagan Center for the Study of Life in the Universe was working with over sixty scientists.[71]

The move into astrobiology is partially due to money concerns. Without congressional action, the SETI Institute cannot receive federal grants for SETI, much less see a new NASA project like the one Senator Bryan killed. Astrobiology and planetary science, however, are politically safe, since they lie within NASA's purview. As a result, the SETI Institute has branched out into these disciplines to win grants that help pay the overhead. These disciplines have the further advantage of expanding the SETI Institute, giving it a larger footprint. Without its Carl Sagan Center, the SETI Institute is reduced to being a telescope and a handful of staff.

Planetary science dates to Galileo and his telescope. Astrobiology, though, is a product of the space age. In 1959 NASA funded its first astrobiology project, and in the 1990s the search for alien life became a mission for NASA. To find alien life on Mars or elsewhere, humans must be open-minded when coming across unfamiliar life-forms. Astrobiology can be described as preparing for the unfamiliar alien through study of Earth's strange life. Astrobiologists study life in exotic places: deserts, ice cores, deep ocean vents, hot springs, high Andean

lakes. This curiosity will, it is hoped, allow robots or astronauts to recognize novel life-forms in other worlds. Yet although alien life has not
been found, astrobiologists take for granted that "life is not a freak
phenomenon confined to Earth, but a widespread and inevitable outcome of physical laws."[72] If nothing else, astrobiologists have expanded
knowledge of Earth. According to one estimate (2006), there are one
thousand astrobiologists in the United States alone.[73]

The SETI Institute realizes it cannot be an astronomical cloister. It
engages the public, tacitly admitting that it expects contact to come
later than sooner. A key component of the SETI Institute's outreach
is to raise science literacy and to prepare the next generation of SETI
researchers. The institute hosts a summer internship for high school-
ers to study astrobiology.[74] Weekly lectures are available on YouTube
and the Internet. The electronic media also boast the institute's weekly
one-hour radio show, *Big Picture Science*.[75] For astronomy buffs who
want to combine learning and vacationing, the institute conducts tours
to places of astronomical interest. In 2012, Seth Shostak led a two-week
tour to Australia to see the total eclipse of the sun.

In August 2010 the SETI Institute held its three-day SETIcon con-
vention in Santa Clara, California, and two years later SETIcon II took
place. According to the institute, these events celebrated "science and
the arts for the public." To the man in the street, SETIcon I and II may
seem like carnivals for science nerds. Prominent scientists spoke at the
panels and lectures, as did science-fiction authors, astronauts, businessmen, artists, musicians, and actors from the *Star Trek* series. Panel
topics included the whimsical and the serious: from Hollywood's depiction of aliens to NASA's Kepler Space Telescope. Media star Professor Alex Filippenko of UC Berkeley got attention when he denied God
was necessary to begin the Big Bang. Auctions were held to benefit the
SETI Institute. Frank Drake and Douglas Vakoch signed autographs.
About a thousand attended the first SETIcon.[76]

The SETIcons were fun events. By contrast, the SETI Institute was
quite serious when it launched SETIQuest and SETILive to involve "citizen scientists" in the search. Jill Tarter proposed SETIQuest in 2009,
when she received the TED Prize, given annually to visionaries in the
arts and technology (other recipients include Al Gore, Bono, Jane

Goodall, and Bill Gates). Upon receiving the award, recipients express their plan to "change the world." Tarter wished to "empower Earthlings everywhere to become active participants in the ultimate search for cosmic company."[77]

SETIQuest was born when the SETI Institute open-sourced SonATA, its operating software for the Allen Telescope Array, and invited volunteer programmers to improve it. The less technically savvy were invited to parse data from the ATA and look for anomalous patterns, which algorithms did not catch. According to the SETI Institute website of April 23, 2010, "a community . . . a tribe to actively involve the world in the ultimate search" was being created.[78] In 2012 the SETI Institute website confessed to failure: a vibrant online community never materialized. The institute had underestimated SETIQuest's funding and resources. SETIQuest was relegated to "a maintenance mode," open to volunteers willing to turn it around.[79]

The public was invited to focus instead on SETILive, another "citizen scientist" project. Less ambitious than SETIQuest, SETILive did not involve the writing and/or tweaking of software. SETILive tried instead to spot possible alien signals in frequencies crowded with human signals. Unlike SETI@home, whose data can be months old, SETILive examined data from the ATA in real time. Users had ninety seconds to examine a "waterfall" image before the telescope moved to another star or frequency. If enough SETILive users spotted a likely signal, the telescope returned for a second look within three minutes.[80] SETILive turned out to be another fiasco. In October 2014 the SETI Institute shut it down, finding that it was too costly.[81]

Although continuing the search and branching into new directions, the SETI Institute has been financially insecure. Its website on August 4, 2015, carried an open letter, "The SETI Institute Needs Your Help," from Bill Diamond, president and CEO, saying that the Institute needs additional funds to protect its current programs.

The SETI Institute has not been alone. SETI@home has also been financially troubled. Andrew Siemion, the SETI director at UC Berkeley, has said that SETI@home has struggled to survive.[82] According to Frank Drake, in recent years the total worldwide spending on SETI has been half a million dollars, all from private gifts.[83]

Major donors have been few and sporadic. The funding of SETI brings no quick gratification, for contact may occur in the distant future, if at all. Furthermore, the scary science-fiction scenario always looms; there is no guarantee that contact will enhance humanity. The lack of stable funding not only retards SETI searches but also makes full-time SETI work a risky career choice for the young. Short of money, the SETI enterprise was giving the impression of declining.

Yuri Milner: A New Era Begins

Suddenly, on July 20, 2015, SETI enthusiasts heard great news. Yuri Milner (b. 1961), a Russian billionaire and Silicon Valley tycoon, announced his Breakthrough Listen project, the donation of $100 million for SETI research. Frank Drake gushed: "It's just a miracle."[84] This is Milner's second grand act of philanthropy. In 2012, along with Facebook founder Mark Zuckerberg (b. 1984) and Google co-founder Sergey Brin (b. 1973), he created the Breakthrough Foundation, whose $3 million awards for achievement in theoretical physics, mathematics, and life sciences overshadow the long-established Nobel Prizes.

Yuri Milner was born in Moscow and majored in theoretical physics at Moscow State University. He has since claimed he was not smart enough to stay in theoretical physics. Turning to business, Milner came to the United States in 1990 for a MBA at the Wharton School of Business, the first non-émigré Russian to do so. Milner has overachieved in his second career choice. He has shrewdly chosen promising business ventures, investing in new Silicon Valley firms, among them Facebook and Twitter. With a net worth of more than $3 billion, he has appeared on lists of very prominent businesspeople. Milner has had a nose for winners and obviously hopes that SETI is another winner. Contact will not make him richer, but it will make him forever famous.[85]

Milner made sure his donation would get maximum publicity. Announcing it on July 20, the anniversary of the Apollo 11 moon landing in 1969, was no accident. Although the new SETI project will be conducted mostly in the United States, Milner told the world at the Royal Society in London, in the company of Stephen Hawking no less, who gave his full endorsement. Also present were Sir Martin Rees,

Britain's Astronomer Royal and Cambridge don; Geoff Marcy, astronomy professor at UC Berkeley and foremost discoverer of exoplanets; Ann Druyan, Carl Sagan's widow and collaborator; Pete Worden (b. 1949), retired director of NASA's Ames Research Center, who is the new chairman of the Breakthrough Foundation; and Frank Drake, who started it all in 1960. These heavyweights guaranteed widespread press coverage.[86]

What will the $100 million do? It stops SETI's decline and gives new life. The money represents far more than the amount currently spent worldwide on SETI. Breakthrough Listen will cover an area ten times greater than earlier programs, one hundred times faster, and with fifty times more sensitivity. And as the abortive NASA SETI project had planned, it will scan the entire 1-to-10 GHz frequency range.[87]

UC Berkeley will get a good chunk of the money, a third of it earmarked for Berkeley's laboratories to build specialized electronics to gather and analyze data. Another third is earmarked for hiring astronomers and, crucially, training the next generation of SETI scientists.[88] Dan Werthimer, already noted for pioneering and running SETI@home, will be in charge of gathering and analyzing data.

The rest of Milner's largesse buys dedicated observation time, something SETI was very much lacking in recent years. According to a contract with the Breakthrough Foundation, the one-hundred-meter Robert C. Byrd Green Bank Telescope, the world's largest fully steerable radio telescope, will receive $2 million per year for five years and in return devote 20 percent of its observing time to SETI.[89] This contract is good not only for SETI but also for the Green Bank Telescope, which was in dire need of a new cash stream. Government cutbacks in science and new priorities were forcing the telescope's owner, the National Science Foundation, to reduce its support, although the telescope is relatively new, completed in 2001.[90]

The Breakthrough Foundation has also contracted with the Commonwealth Scientific and Industrial Research Organisation, Australia's science agency, to use the Parkes radio telescope. Starting in July 2016, Parkes will devote 25 percent of its telescope time in the next five years to SETI.[91] As with the Green Bank Telescope, the deal represents a welcome infusion of cash to the aging Parkes telescope, built in 1961.

Like the short-lived NASA SETI program, Breakthrough Listen will have two search strategies: targeted and all-sky. The targeted search will focus on the nearest one million star systems, hoping to catch leakage signals, in case ET is not trying to contact other civilizations. As for Earth's immediate neighborhood (by astronomical standards), the telescopes will examine the closest one thousand stars with great precision, hearing ET's equivalent of an ordinary air traffic control radar. According to Geoff Marcy, the increased telescope time will allow "longer dwell times on the targets" to sense "weaker, fainter radio signals."[92]

On the other hand, the all-sky strategy might work. Kardashev might be right. There may be Type II civilizations in the far reaches of the Milky Way, and beyond our galaxy, Type III civilizations may be harnessing the energy resources of their galaxies. If supercivilizations exist, they might be curious and transmitting extremely strong signals. A wide sky sweep might succeed. Breakthrough will sweep the 100 billion star systems at the center of the galaxy and will look at 100 of the nearest galaxies.[93]

SETI people are well aware that a communicating ET may not be using radio. To cover this possibility, Breakthrough Listen plans an optical search, using the optical telescope at the Lick Observatory in San Jose, California, to search for continuous laser signals. This optical search will be so powerful that it can detect a 100-watt laser from the nearest star system.[94]

Collecting data is only part of the search. Raw data need careful analysis. After all, ET, strange and mysterious, is not expected to make it easy, broadcasting in English on the BBC world network. To complicate matters, record quantities of data will be available. Breakthrough Listen will rely on SETI@home's volunteers, but they will not be enough. New software for data crunching will need to be developed. Breakthrough Listen will also reach out to other researchers by making its software and hardware compatible with other telescopes. This transparency not only widens the search but also, according to Milner, is necessary to forestall the "many conspiracy theorists."[95]

Another side to SETI has been messaging to extraterrestrials. Milner's grand design also includes a Breakthrough Message competition. People around the world will be tempted with $1 million in prizes to compose messages for ET. Yet, Milner must be aware of the controversy surrounding Active SETI, for he denies plans to broadcast any of the messages the competition produces. Breakthrough Message may be a scheme to publicize the search.[96]

Why is Yuri Milner spending $100 million on SETI? He has pointed to his scientific background, and even to his name. He was born in the year of Yuri Gagarin's ascent into space, and his patriotic parents named him after the Russian space pioneer. In his childhood he was fascinated by the works of Carl Sagan and Isaac Asimov. Yet biography is insufficient. Milner notes that the possibility of extraterrestrial life has always intrigued humanity, and he adds that Silicon Valley can process far more data than ever. In short, more than any time in the past, success is possible. Milner does concede that ten years of searching may result in failure, and if failure results, he promises to fund another ten years.[97]

What Milner has not said is noteworthy. Unlike some of the SETI pioneers, he does not mention the benefits accruing to humanity from contact: no references to cancer cures, immortality, or world peace. Milner does not even claim the universe is teeming with life. He does believe, though, that if humanity is alone, "it would be such a waste of real estate." All he wants to do is "to help find an answer."[98]

This prosaic language suggests that the optimism and the utopian yearnings of SETI's early years are gone—or maybe not. Milner the canny businessman may be keeping his true feelings to himself. In hoping to discover strange beings and places, he has placed himself in the same continuum as the explorers of old. Yuri Milner may be a dreamer after all.

All that remains is for ET to call.

Conclusion

Reasons abound why SETI could be a waste of time and money. Life may be confined to Earth. If life exists elsewhere, it may be primitive, no more advanced than bacteria; if advanced, it may not be self-conscious; if it is self-conscious, it may not be technological—after all, Neanderthals did not have telescopes. If extraterrestrial life has the technology to communicate across the stars, it may be solitary. The Galactic Club may be a fantasy of lonely Earthlings who have overindulged in science fiction. Finally, it could be that extraterrestrials are curious, but they are not sending microwave transmissions. They reach other worlds through gamma rays, neutrinos, telepathy, or a medium unknown even to the imagination of science fiction.

The odds are imposing. Yet individuals do overcome the odds and win the big lottery. SETI might also win its lottery, when it turns out the galaxy does contain advanced extraterrestrials who have language and whose microwave signals reflect that language. When Earth receives these signals, SETI will have proven the skeptics wrong.

Contact

What happens next? Though the future is always murky, the near future, if placed within the lifetimes of today's people, is recognizable. When Albert Einstein died a half century ago, he knew of jet aircraft, television, telephones, and nuclear power. He lived close enough to the twentieth-first century to probably adjust to new technologies. By contrast, Abraham Lincoln knew little of twenty-first-century technology. Living in the age of candles and carriages, Lincoln would be technologically lost in our time. Barring unforeseen natural or political cataclysms, we can project our world to the 2050s or 2060s. The next century is too unfamiliar to discuss humanity's reaction to contact.

The discovery of extraterrestrials need not take the SETI route. The wreckage of an alien vessel may be found. Perhaps, an interstellar phenomenon or process is recognized as artificial, the work of alien intelligence. Chlorofluorocarbons are an example. Not found in nature, if found beyond Earth, they indicate a technological mind. As befits a SETI study, our contact scenario will result from a radio signal, although a laser transmission is possible.[1]

A non-SETI person could receive and identify the alien signal. If serendipity should smile on a non-SETI astronomer, the SETI community would lose the fame it deserves. If life is fair, either the ATA, SETI@home, SERENDIP, the SETI League, even Argentine, Australian, or Italian SETI should get the signal. If the signal should come during the time Yuri Milner is funding SETI, he would like one of his projects to receive it. But life is not always fair.

According to the contact protocols, the fortunate recipient should contact other astronomers to verify the signal. If telescopes around the world receive a continuous signal, dismissing a terrestrial source is rather easy. On the other hand, if the signal is non-terrestrial but of natural origin, dismissing it may prove challenging.[2] When British astronomers discovered pulsars in 1967, they briefly considered the possibility of "little green men" sending coded signals. They eventually realized that the regular pulses of electromagnetic radiation were coming from rotating neutron stars, or pulsars. Pulsars were simply part of the unknown waiting to be discovered and enter the astronomy corpus.[3]

Once astronomers cannot explain away the signal and believe an intelligence sent it, the contact protocols mandate that they contact the United Nations, whose secretary-general will inform the world. This scenario assumes that an astronomer in the know resists the temptation of instant fame and does not hold a press conference. It may be that reporters find out before an official announcement and a media frenzy begins. The media aside, a politician who wants to preen before the world's cameras might preempt the secretary-general.[4] The news should not shock the public. Ufology and science fiction have long conditioned it to accept alien minds. In a 2005 poll, 60 percent of Americans reported their belief in intelligent extraterrestrial life.[5]

The radio signal will probably be a narrow band of few frequencies, as SETI astronomers expect. Wideband signals are more likely distorted by free electrons and protons in space, but they carry more frequencies and information. Which one will ET use? As far as Earth is concerned, the medium does not matter; the message that the signal carries is all-important.

Seth Shostak notes that most SETI instruments average the radio noise to make a possible signal stand out. Unfortunately, this process wipes out the message, leaving knowledge of a signal but not the message itself. SETI does this because its priority is detecting ET. Only later, if ET has continued to transmit, giving decoders a large stock to examine, will the focus shift to extracting information.[6] The easiest message to decipher would be coded in a human language. This means, though, that ET has observed Earth, a situation that might be disturbing. If the message is not quickly deciphered, a waiting period begins, which will last months, years, or perhaps forever.

The Waiting Period

Citing the public good, politicians will likely attempt to dominate the waiting period. They will ignore the SETI Protocols, bypass the United Nations, and pursue their perception of the national interest. They will determine the increased spending on new equipment for more sensitive reception. In addition, they will control the horde of newly hired cryptographers and demand instant news of a decoding.[7] One should never forget that physicists built the first atomic bombs, but politicians decided whether to use them. In his SETI novel, *Contact*, Carl Sagan anticipated the scientists' loss of control. In compensation, science, especially astronomy, will be showered with money.[8]

As for the SETI community, its leading lights will receive the proverbial fifteen minutes of fame. They are the traditional experts, and the media will besiege them for information on all matters SETI as well as speculations on extraterrestrials. This attention will fade. Other astronomers, as well as science-fiction writers, physical and social scientists, indeed, anyone who projects intelligence, will pontificate. With new "experts" emerging, the SETI people will have to share the stage, if not fight to stay on it.

Whether to respond will be hotly debated. Although the signal's message may not have been decoded, a reply can be beamed to the star system from which the signal originated. Even if the extraterrestrials make no sense of the human reply, they will know Earth has received their transmission.

The debate over responding will be short. Freelancers will decide the issue by bombarding ET's star system with responses. Even though ET's message has not been deciphered, freelancers will hope ET understands their responses, which will range from the silly to the sensible. Will this be bad? SETI enthusiasts assume that extraterrestrials speak in a common voice. Unlike benighted Earthlings, they have their version of world government. Yet, a united extraterrestrial society may be a projection of our utopian dreams. As Freeman Dyson said, the babel of voices we transmit will give an honest picture of humanity.[9]

Decoding ET's message may take time. Although SETI optimism holds that science and mathematics are universal languages, extraterrestrial perception of the natural world may differ radically from ours, and the belief in congruence may be wishful thinking.[10] Earth's history is somewhat discouraging. The discovery of the Rosetta Stone in 1799 did not quickly result in the translation of Egyptian hieroglyphs. The Rosetta Stone, circa 200 B.C., contained a pharaonic decree in both Greek and hieroglyphs. Although scholars could easily read the Greek, they could not make sense of the hieroglyphs' pictures until Jean-François Champollion (d. 1832) realized that hieroglyphs consisted of both phonetic and ideographic symbols. Although later generations of scholars deciphered other ancient writings, they are still befuddled by Linear A of the Cretan Minoan Civilization and by the mysterious script of the Indus Valley Civilization.[11] If human script can stump scholars, an extraterrestrial missive might be impossible to decipher.

ET would help by sending hints on decoding. We can only hope that these "hints" are as obvious to us as they are to ET.

Conspiracy Theories

Cryptography and universal language—these refer to the establishment view of contact, the stuff of the *New York Times* and *Scientific American*. However, other views of contact will be bubbling in the social media, in the popular press, and on the street.[12] The waiting period will produce deniers. The same mind-set that refuses to believe the 1969 moon landing will also deny an alien signal. Cynics will give all sorts of reasons why contact is a hoax. They will accuse governments of distracting from the politics of the day, astronomers of wishing to increase funding, and SETI people of seeking publicity. Deniers will have a nearly endless list of conspiracies to choose from.

Ufology, despised and dismissed in establishment circles before contact, will be a big winner. With the United Nations in tandem with world leaders and astronomers all admitting that extraterrestrials exist, ufologists will claim validation. They will smugly announce they knew all along and were ahead of the curve.[13]

Ufologists have seen cover-ups in the past and will not stop now. Insisting the waiting period is bogus, they will claim the government has decoded the alien message and is hiding it. As proof, they will point to new contactees who allegedly know the message the authorities refuse to reveal. Will the public buy into this counternarrative? This much is certain: accusations of official cover-ups will occur in a post-contact era when ufology will be stronger.

While ufologists and conspiracy theorists preen themselves, the rest of humanity will learn of the Active SETI debate over the wisdom of communicating with ET. That debate, once the province of the small world of astronomy, will become mainstream. However, in one sense, the post-contact debate will be different. In addition to discussing the wisdom of transmitting to extraterrestrials, it will also include the unofficial replies already sent to ET. The big question will be whether humanity should worry.

Human Reactions

The transmitting world's distance from Earth will determine human reactions. If the distance is within a human lifetime—within twenty light-years, for example—a number of people alive today can receive a reply to responses beamed in ET's direction, and they can hope this reply will be easier to decode and reassuring. On the other hand, if ET lives several hundred light-years away, the span of many lifetimes, the great majority will lose interest after the novelty of receiving the signal wanes. Human presentism will assert itself.

A short waiting period might be unsettling if the known information of the transmitting planet results in an unpleasant portrait of ET. It is already possible to know the size of an exoplanet and whether it is gaseous or rocky. The James Webb Space Telescope, scheduled for launching in 2018, will be able to identify the spectrographic signature of water vapor. At the time of contact, astronomers should have more sophisticated tracking instruments to assay the atmosphere, temperature, and topography of ET's world. This information will lead to speculations on the appearance, size, and behavior of the planet's inhabitants.

Artists will draw portraits of ET that will range from the benign to the hostile. As for ET's behavior, Percival Lowell showed that limited knowledge is no bar. His Martian pontifications were taken seriously, at least by much of the public. There is no escaping that the more ET's world physically differs from Earth, the stranger will be ET's appearance. Some portraits of ET, especially if ET's world lacks water and/or oxygen, will be gruesome. ET's bizarre body will make humans uneasy, especially if ET's planet is nearby.

Finally, the possibility looms that the message will never being deciphered. The longer the signal remains an enigma, the more it becomes a dial tone, simply the knowledge that something is out there. The future will be a nuanced version of today. Ufology will continue to seduce. Astronomers and biologists will titillate the public with imagined scenarios of ET's world, much as Lowell and Bracewell did in their day.

ET's Message Decoded

If the message should be decoded, how will humanity react? The answer depends on the content of the message and the distance of the transmitting civilization. At one extreme, Earth receives a short, innocuous greeting, sent from a great distance. Whether the extraterrestrials are hostile remains unknown, but distance (one hopes) protects Earth, and as a result these extraterrestrials pose little threat. Only in the distant future—say, thousands of years—will a response to our reply arrive. One effect of this brief message from afar will be a debate whether to expend resources to search for closer intelligent worlds.

A brief, innocuous message from nearby—defined as the alien reply arriving within a lifetime—is something else. A debate will begin whether to respond to the message. The pros and cons of the Active SETI debates would be rehashed with the added complication of our knowledge of ET's physical world and projected portraits. However, while the "official" organs debate, independents, if possible, will preempt and transmit their own responses.

The extreme opposite to a brief greeting is receiving the Galactic Library. Another coup for SETI optimists! Its presumptions about a Galactic Club have not been wishful thinking. The significance of receiving this text also depends on distance. If all the Galactic Club members live hundreds, if not thousands, of light-years distant, they are too far for a conversation. Their replies to our queries will arrive many centuries in the future and will concern matters of minor interest to our descendants. Only the arrogance of presentism makes us believe our descendants will share our concerns. After all, Isaac Newton would not have asked about string theory or a cure for Ebola. He probably would have inquired about his other interests: alchemy and witchcraft. The Galactic Library of distant beings is akin to an ancient manuscript with long-dead authors who cannot be questioned, creating the challenge of divining their thoughts. On the other hand, if at least one member of the Galactic Club lives close enough for an intragenerational conversation, there is a chance of receiving timely answers to questions we deem important.

In the best of scenarios, philosopher kings rule in the Galactic Club. The Galactic Library gives information to plug into our existing structures as well as leads on new avenues of endeavor. Earth profits from the experiences of older, more advanced civilizations. We learn how other civilizations handle mega-crises. We face the future with more assurance than previous generations.

Maybe not! The consequences of contact, good or bad, like all the future, are unknown. SETI advocates concede that the short-run consequences of contact may be unsettling. When confronted with the galaxy's superior civilizations, many humans may suffer disorientation and lose self-esteem. Experts in many fields will face the shock of having their years of study, their learning and wisdom, devalued.

SETI optimists are unfazed. Even if humanity requires several or more generations to adjust, the optimists take the long view and see human welfare increasing. This positive outlook permits SETI advocates to blithely consign future humans to a temporary unhappiness.[14] Nonetheless, is this optimistic prognosis valid? Will a better life follow the disorientation? We could be dealing here with a variety of the twentieth-century political disease: sacrificing a generation or more for a rosy future that unfortunately remains forever on paper.

So far we have been assuming that the Galactic Library contains a treasure trove of valuable information. In reality, though, it may disappoint. If extraterrestrials perceive their environment through different senses—for example, they smell as opposed to seeing and touching, or they possess nonhuman senses such as sonar—their "science" may be baffling. How much of it will be understood is a question only the future can answer.

Another challenge may be appreciating ET's culture. The fine arts, literature, and even the social sciences rest on human emotions and senses. Blind extraterrestrials will not paint a *Mona Lisa*, nor will the deaf produce a Mozart. Our literature in large part reflects human vices and virtues. Will ET's different body chemistry allow the human mix of capital sins and cardinal virtues? Will ET's literature deal with lust and envy as well as justice and courage? Could ET understand Homer? We moderns find the characters in the *Iliad* and their obsessions with honor a little strange.[15]

Religion from the Stars

A SETI truism sees humanity learning how other civilizations flourished and died. If humanity is like other forms of galactic life, an accumulation of star dust, the history of these older civilizations should give a glimpse into humanity's future and destiny. The stars will give answers that philosophy and religion once gave.

Several SETI people look for religious guidance from the stars. Jill Tarter, Carl Sagan, and Steven Dick see little use in earthly religions, blaming them for wars and ignorance. Jill Tarter has said that if extraterrestrials have religion, it is universal and respects science. Dick sees a cosmotheology emerging in the future, a religion that incorporates cosmic evolution as well as science. If cosmotheology has a god, it will be a natural god who leaves lower creatures alone.[16]

Yet traditional religions have great attractions. Apart from the promise of immortality, there is the personal element of the here-and-now. Chet Raymo (b. 1936), author and professor emeritus of physics at Stonehill College, has rejected religion yet concedes its attractions: "Religion . . . provides a sense of belonging to a group, a history, and a culture in which to take pride, great works of art, stirring literature, service to the poor and needy, satisfying liturgical celebrations of creation, rites of passage."[17]

Finding substitutes for religion's appeals has traditionally challenged atheism. Apart from the personal element, atheism also lacks the supernatural to set and enforce rules. An anecdote from World War II that may be apocryphal illustrates the atheist conundrum. An elderly Jew about to be shot by a Nazi soldier said to his executioner, "God is watching." A god-free universe has no God to punish the Nazi, who may have survived the war, prospered in the new Germany, and died peacefully in old age, unrepentant to the end. Life is not fair. If there is no god, Stalin and Mao will never be punished for slaughtering millions. Although history judges them badly, they are too dead to know and care.

To compete with religion, unbelief needs an ethical system. Secular humanism does provide a self-contained solution. Accepting that life is an accident and the universe lacks intrinsic meaning, it exhorts humans to create ethics from the best of human experience. The self-interest of humanity demands that individuals respect each other and create a just society.

SETI has its own solution to the need for ethics. The wisdom of superior extraterrestrials will guide humanity. A moral order from the stars will give individuals a greater purpose to their lives than secular systems that range from humanism to hedonism. Humans will join the great cosmic journey of the universe. This millennial religion of tomorrow will provide answers to questions we cannot imagine. In addition, a crucial difference exists between this religion and the god or gods of traditional religions who remain unseen: extraterrestrials will eventually show themselves. In short, extraterrestrials will succeed today's gods and will perform their own version of the Last Judgment. What else did Carl Sagan mean when he wrote: "There may even come a day when we shall be called to account for our stewardship of the Solar System."[18]

Contact and the Threat to Religion

Contact with extraterrestrials has been seen as threatening to Christianity, but less to other religions. According to the Dalai Lama, Buddhism easily accommodates extraterrestrials.[19] The Koran refers to extraterrestrials. On the other hand, the Bible does not mention them, and it makes humanity the center of God's attention. Can the fall from grace in the Garden of Eden, the Incarnation, and the Crucifixion fit into a biological universe? Did Christ replicate his death on other worlds? Arthur C. Clarke believed that contact will weaken Christianity, and this perception has often been echoed.[20]

British theologian David Wilkinson is unfazed by the so-called de-centering argument. It is a misreading of Christianity to see the human race in the center, he says. "God is the centre of all things and we are creatures given status by his love." With these words, Wilkinson accepts extraterrestrials, although their exact status in God's scheme of things remains an open question. Whether the Christian rank-and-file will share Wilkinson's equanimity after contact also remains to be seen.[21]

Surveys show that the non-religious more are likely to believe in extraterrestrials than church people are.[22] It is distinctly possible that nonbelievers are prone to transferring a residual religious instinct into the material world of the cosmos. In contrast, churchgoers do not need succor from extraterrestrials. They have their god.

The question of Christianity's reaction to contact came up at SETI-Con II in 2012, which devoted a session to this topic. Douglas Vakoch noticed the large audience and remarked that the topic of religion and extraterrestrials invariably draws well. The panelists—Vakoch, Shostak, John Gertz, and science-fiction author Robert Sawyer (b. 1960)—were noncommittal, though, on contact's effect on Christianity. They noted that contact does not worry theologians from the major faiths.[23]

The director of the Vatican Observatory, Jesuit father and astronomer José Gabriel Funes, has said that aliens may be a different life-form and do not need redemption.[24] Catholic theologian Thomas F. O'Meara is instead open to "alternate salvation histories" that correspond to other intelligent creatures. He even concedes the possibility of some extraterrestrial races having no notion of an afterlife and being quite happy.[25] One could assume that with the Galileo fiasco in mind, the Roman Catholic Church is being very open-minded this time.

Contact should not be compared to Christianity's shock when Darwin introduced the theory of evolution. Christianity was overly invested in the Genesis creation story. Rejecting it naturally caused a crisis, which for many Christians is still real. On the other hand, ET is different from Adam and Eve. Most Christians have invested little emotional and intellectual capital in the nonexistence of extraterrestrials.

The exception to this openness may come from fundamentalist Christians who cannot find extraterrestrials in the Bible.[26] If a signal is untranslated, fundamentalists may be among the deniers and conspiracy-minded. Yet, can denial continue forever? Shostak believes it cannot. Besides, Christianity has survived by adjusting. At SETICon II, Robert Sawyer mentioned a Baptist minister in Georgia who challenges evolution but is open to extraterrestrials. The minister uses the analogy of a couple that decides to have a second child and loves it as much as the first.

Contact's Impact on the SETI People

Those most invested in contact and dialogue with ET are the SETI people and the New Age enthusiasts, who believe contact will better humanity. To their shock, ET might not live up to their expectations. ET's culture might not be science-based, as SETI people like to believe. In ET's world, scientists might be lowly artisans, much like medieval artists who ranked below warriors and clerics in the pecking order. ET might not send a scientific message, and might not fathom one that is received. We know from human experience that not all societies would fashion the same message. If the Islamic Republic of Iran launched its version of Sagan and Drake's *Pioneer*, instead of nude figures the Iranian Pioneer would carry a religious text: "Allahu Akbar" at the very least. Although Iranian scientists developed the technology, the ruling clergy would compose the message.

Perhaps ET values science and religion less than good cheer. Instead of transmitting a Galactic Library, which contains a perpetual energy source or a reconciliation of quantum mechanics with Einstein's relativity, ET transmits a joke collection and requests samples of earthly humor. Humor from outer space does seem preposterous, but defining ET's message as scientific before knowing ET is also preposterous.[27]

Let's assume the contents of the Galactic Library are serious tomes, at least by scientific standards. However, the solutions that extraterrestrials have devised for the challenges of their societies may not fit into our world. Digesting extraterrestrial knowledge and advice cannot be compared to American automakers adopting Japanese industrial practices, plugging "just in time" production into an existing framework.

Extraterrestrials have evolved in worlds with unique chemistries, temperatures, and geological histories. It is very unlikely, as Shostak once noted, that ET has a humanoid shape. This has implications for the SETI narrative. Sagan famously believed that advanced extraterrestrials had survived technological adolescence and learned to live in harmony. Furthermore, Sagan hoped, extraterrestrials had the missionary impulse to share their survival strategy. Unfortunately for this SETI dream, ET's cultural "wisdom" will impress humans only if they relate to ET's appearance. If ET breathes ammonia and resembles a roach, humans will need an enormous effort to relate to this life-form and its "wisdom."

If ET's appearance is no bar, will humanity approve of ET's advice? The neurotransmitters of human bodies will differ from the neurotransmitters of extraterrestrials. Their behavior will not match ours. Their solutions to problems we share might be odious. Consider overpopulation. Humans contain it through abstinence, birth control, abortion, or infanticide. Extraterrestrials might instead cull the herd, randomly choosing, every so often, a certain number to die. Their bodies are a feedstock for transplant surgery, allowing extraterrestrials to live long, provided they are lucky enough to survive the cull.

Another depressing revelation might be ET's advice for surviving lethal technologies. Extraterrestrials might survive by killing their "maverick" scientists; they execute their Einsteins for even suggesting the splitting of the atom. In ET's world, dabbling in "the dangerous arts" such as mixing the genes of different species, creating noxious chemicals, and threatening the environment leads to painful death. A secular inquisition protects ET's future. In outer space, the fate of Giordano Bruno (1548–1600), the freethinking astronomer and philosopher whom the Inquisition executed, is the norm. With science and technology having reached a near dead end, the only change among extraterrestrials is cultural. Hence, their curiosity in our ways. The SETI community must ignore this scenario of stasis; it must believe in progressive extraterrestrials, or else the search enterprise is somewhat pointless.

Although science fiction has presented countless scenarios of ET's behavior, in all truth, no one knows the real ET. This much is sure: extraterrestrials will have evolved in a non-Earthlike world and perforce will be different, probably strange, and perhaps repulsive. The scientists and laypeople who have invested emotional and intellectual capital into extraterrestrials may experience the greater shock and disorientation. In short, the salient question about contact is not its effect on religion but rather its effect on both the SETI community and the New Age enthusiasts who expect a meaningful denouement.

Is SETI Worth the Effort?

Contact is no certainty. Furthermore, the odds of favorable results are slim. Believing in eventual benefits is an act of faith. A pessimist would eschew the SETI enterprise, or at least oppose Active SETI. David Brin believes that instead of "rushing toward contact," we should be "preparing our heirs to be ready for it."[28]

If ET exists, serendipity might eventually lead to contact. NASA stands as good a chance of discovery as any. After all, one of its missions is searching for life. Instead of finding alien bacteria, NASA might stumble onto non-radio signs of ET. The passive SETI effort might as well be made; it merely hastens a future that might arrive. Besides, the SETI community has prepared us for contact. It has provided detection protocols and has debated the important question of contact's merits. The future is always uncertain, and it never unravels fully as expected, yet without preparation that walk into the unknown is less steady. Moreover, SETI apologists have countered ufology. They have rationally focused attention on extraterrestrials.

The SETI enterprise is a search for knowledge. Even if contact never comes, SETI has spawned new technology. As for learning that ET will share, it should be noted that information should not be confused with wisdom. A search for wisdom will invariably set humanity up for disappointment. If knowledge is instead seen as good in itself and its expansion benefiting humanity, gaining new information best justifies SETI.[29]

Should the U.S. government fund SETI? The answer is no. As a private endeavor, SETI does not compromise government. Even a cent of public money spent on SETI might be construed as a government endorsement of ufology. Besides, government chooses from many competing interests. SETI's problematic outcome makes it a tempting target. If Senator Bryan had not killed NASA SETI, another politician would have. SETI is safer in the private sector than under government's uncertain embrace. In any case, even today's limited SETI might succeed; after all, Columbus's voyage cost little. If privatization means prolonging the date of contact, that is not bad. Humanity can wait.

Humanity prepares for contact by contemplating possible scenarios. Once contact takes place, however, pre-contact speculation will need updating. The real world of contact will rule. The Age of Discovery provides a useful analogy. Europeans initially saw the Americas in terms of their culture. The Roman Catholic Church in Spain hoped to find pagans to convert, and these they found; the conquistadores hoped to find gold, and this they did. All the same, the Spanish never foresaw the enormous African slave trade, nor the impact of the Columbian Exchange on the world diet, nor the worldwide inflationary shock of America's precious metals. One thing is for sure. Contact with extraterrestrials begins an adventure, a step into the dark. Whether the future will hail or revile the astronomers responsible, only time will tell.

NOTES

Introduction

1. In a 1975 article in *Scientific American,* Carl Sagan and Frank Drake wrote that contact with extraterrestrials will enrich humanity ("Search for Extraterrestrial Intelligence"). Allen Tough's *When SETI Succeeds* contains a variety of scenarios of contact's salutary effects. The use of the term "contact" for humanity's first alien encounter dates from 1945, in science-fiction writer Murray Leinster's short story "First Contact."

2. Cocconi and Morrison, "Searching for Interstellar Communications."

3. Drake tells his story in Drake and Sobel, *Is Anyone Out There?*

4. Billingham et al., *Social Implications*, 44.

5. Billingham interview in Swift, *SETI Pioneers*, 256.

6. Kilgore, *Astrofuturism*, 150.

Chapter 1. Before SETI: The Age of Pluralism

1. Bailery, *The Greek Atomists and Epicurus,* 1–3; Hicks, *Stoic and Epicurean,* 163–65; Dick, *Plurality of Worlds,* 7–10, 19.

2. Dick, *Plurality of Worlds,* 12–16.

3. Guthke, *The Last Frontier,* 3–4, 33–36; Dick, *Plurality of Worlds,* 9–10.

4. Crowe, *The Extraterrestrial Life Debate,* 10–37.

5. Ibid., 161; Guthke, *The Last Frontier,* 98–105; Dick, *Plurality of Worlds,* 185–86.

6. Mumford, *The Story of Utopias,* 59–78.

7. Crowe, *The Extraterrestrial Life Debate,* 216–17, 231–32.

8. Guthke, *The Last Frontier,* 325–26.

9. Oliver and Billingham, *Project Cyclops*, 13–14.

10. See Larson, *Evolution.*

11. Hoyt, *Lowell and Mars,* 5–7, 200.

12. Strauss, *Percival Lowell,* 3–10.

13. Webb, *Where Is Everybody?* 38; Strauss, *Percival Lowell,* 200.

14. Strauss, *Percival Lowell,* 154–59.

15. Lowell, *Mars as the Abode of Life,* 142–44, 205–7; Hoyt, *Lowell and Mars,* 219, 288–89.

16. Hoyt, *Lowell and Mars,* 85–86, 99.

17. Strauss, *Percival Lowell,* 230–32; Hoyt, *Lowell and Mars,* 168, 179.

18. Hoyt, *Lowell and Mars,* 202–3; Guthke, *The Last Frontier,* 369–82; Bergreen, *Voyage to Mars,* 178–79.

19. W. Sullivan, *We Are Not Alone,* 23; Shklovskii, "Is Communication Possible?" 5–6; Asimov, "Of Life Beyond," 45.

Chapter 2. Science Fiction and Ufology: Strange Cousins

1. Murray, *H. G. Wells,* 96; Guthke, *The Last Frontier,* 386. Apart from puncturing Victorian smugness, Wells took aim at religion. In *The War of the Worlds,* a curate joins the protagonist to hide from the Martians. He is such a sniveling coward that the disgusted protagonist kills him.

2. Del Rey, *The World of Science Fiction,* 26; Clareson, *Understanding Contemporary American Science Fiction,* 11–12.

3. Poundstone, *Carl Sagan,* 8–9.

4. Dick, *Life on Other Worlds,* 402.

5. Cantril, *The Invasion from Mars,* is the classic view of the panic. A study that questions the extent of the panic is Robert E. Bartholomew, "The Martian Panic Sixty Years Later: What Have We Learned?" *The Skeptical Inquirer* 22, no. 6 (November/December 1998), http://www.csicop.org/si/show/the_martian_panic_sixty_years_later_what _have_we_learned.

6. Cowan, *Sacred Space,* 7.

7. In addition to *Last and First Men* and *Star Maker,* see Aldiss, *Trillion Year Spree,* 194–98.

8. See Clarke, *Childhood's End.*

9. Baxter, "SETI in Science Fiction," 354–61.

10. Peebles, *Watch the Skies,* 8–35. According to a 2002 Roper poll, 56 percent of Americans believe UFOs are alien, and 72 percent believe the U.S. government hides this fact. Previous polls show similar results. See Shostak, *Confessions of an Alien Hunter,* 112.

11. For Adamski's story see Leslie and Adamski, *Flying Saucers Have Landed.* See also Sieveking, "Forteana," 31; Disch, *The Dreams Our Stuff Is Made Of,* 2; Peebles, *Watch the Skies,* 93–95.

12. For an analysis of Adamski and the other contactees, see Hague, "Before Abduction."

13. Appleyard writes: "Alienology is riddled with a sense of original sin, with a sense not just that we are making a mistake, but that we can only make mistakes" (*Aliens,* 41).

14. Peebles, *Watch the Skies,* 166; Moffitt, *Picturing Extraterrestrials,* 47–58, 331, 338–39.

15. Dean, *Aliens in America,* 6, 34–42.

16. Moffitt, *Picturing Extraterrestrials,* 26–42; Plank, *Emotional Significance of Imaginary Beings,* 101, 135–49; Roland, *Investigating the Unexplained,* 132–33. According to

a 1996 Gallup poll, over 50 percent of men and 40 percent of women believe that extraterrestrials have arrived. See Cowan, *Sacred Space*, 83. See also the Roper Poll cited in note 10 above.

17. Cowan, *Sacred Space,* 99–100. According to Cowan, ufology is "a subcultural zeitgeist that will not go away despite all attempts at debunking, discredit, and denial."

18. Dean, *Aliens in America,* 59–60; Harrison, "The Science and Politics of SETI," 151–52.

19. Ted Peters, "Myth in the Heart of Science: Evolutionary Progress as Myth in Astrobiology and UFOs," March 29, 2012, www.pcts.org/meetings/2012/PCTS2012Apr-Peters-MythSci.pdf.

20. Grinspoon, *Lonely Planets,* 364–71.

21. Oliver and Billingham, *Project Cyclops,* 179–80; Beck, "Extraterrestrial Intelligent Life," 13; Carl Sagan, "The Quest for Extraterrestrial Intelligence," *Cosmic Search* 1, no. 2 (March 1979), http://www.bigear.org/vol1no2/sagan.htm. See also Sagan, *Cosmos,* 314–15.

Chapter 3. The Road to SETI: Fearing the Future

1. Eisenberg, "Philip Morrison—A Profile"; Terkel, *The Good War,* 507; see also Morrison's obituary in the *New York Times,* April 26, 2005, http://www.nytimes.com /2005/04/26/science/26morrison.html?pagewanted=print&position=&_r=0.

2. "A Report to the Secretary of War," *Bulletin of the Atomic Scientists,* May 1, 1946, 2–4; Gilbert, *Redeeming Culture,* 42.

3. Gilbert, *Redeeming Culture,* 48–50; Boyer, *By the Bomb's Early Light,* 51–71.

4. Boyer, *By the Bomb's Early Light,* 72–75, 93–106.

5. Ibid., 352–53.

6. Eugene Rabinowitch, "Science and Party Platforms," *Bulletin of the Atomic Scientists,* 15 (December 1959): 405–9.

7. Morrison, "If the Bomb Gets Out of Hand," 6.

8. Lieberman, *The Strangest Dream,* 103–9; "Robeson, Mann Join New Peace Crusade," *New York Times,* February 1, 1951; "Atomic Bomb Aide Tells of Red Ties," *New York Times,* May 8, 1953; Eisenberg, "Philip Morrison—A Profile," 38.

9. See Morrison and Tsipis, *Reason Enough to Hope.*

10. Wattenberg, *Fewer,* 8, 17, 91–92.

11. U.S. Department of Commerce, United States Census Bureau, *Historical Estimates of World Population,* http://www.census.gov/population/international/data/worldpop/table_history.php.

12. U.S. Department of Commerce, United States Census Bureau, *World Population: 1950–2050,* http://www.census.gov/population/international/data/worldpop/table_population.php.

13. Wattenberg, *Fewer,* 21.

14. According to a 2001 study in *Nature,* world population will peak at 9 billion souls around 2070 and then begin to decline. See Warren Sanderson, Wolfgang Lutz, and Sergei Scherbov, "The End of World Population Growth," *Nature,* August 2, 2001, 543–45.

15. Dyson, "Search for Artificial Stellar Sources of Infrared Radiation"; Schewe, *Maverick Genius,* 139–44.

16. Schewe, *Maverick Genius,* 34, 268–74.

17. W. Sullivan, *We Are Not Alone,* 257–59.

Chapter 4. The 1960s: The Quest Begins

1. C. M. Jansky Jr., "My Brother Karl Jansky and His Discovery of Radio Waves from beyond the Earth," *Cosmic Search* 1, no. 4 (Fall 1979), http://www.bigear.org/vol1no4/jansky.htm; Basalla, *Civilized Life in the Universe,* 133; Michaud, *Contact with Alien Civilizations,* 34.

2. W. Sullivan, *We Are Not Alone,* 198–202.

3. Purcell, "Radioastronomy and Communication through Space," 140–41.

4. Drake and Sobel, *Is Anyone Out There?* 32.

5. W. Sullivan, *We Are Not Alone,* 199.

6. Drake interview in Swift, *SETI Pioneers,* 62, 68.

7. Drake and Sobel, *Is Anyone Out There?* 27.

8. Geirland, "The Quiet Zone," 140–42; Drake interview in Swift, *SETI Pioneers,* 61.

9. Drake recalls Project Ozma in his Swift interview (*SETI Pioneers,* 60–68) and in Drake and Sobel, *Is Anyone Out There?* 27–43.

10. Morrison interview in Swift, *SETI Pioneers,* 22.

11. Drake interview in Swift, *SETI Pioneers,* 57; Drake and Sobel, *Is Anyone Out There?* 5; also Drake's "A Reminiscence of Project Ozma" in *Cosmic Search* 1, no. 1 (January 1979), http://www.bigear.org/vol1no1/ozma.htm.

12. William L. Lawrence, "Radio Astronomers Listen for Signs of Life in Distant Solar Systems," *New York Times,* November 22, 1959.

13. "Earth to 'Listen' for Other World," *New York Times,* April 3, 1960; Drake and Sobel, *Is Anyone Out There?* 31–40; Drake's interview in Swift, *SETI Pioneers,* 64–65.

14. Gerrit Verschuur, e-mail to author, October 11, 2014.

15. Morton holds that Drake is more important than Morrison and Cocconi because Drake "put the idea into practice" ("A Mirror in the Sky," 172).

16. Much has been written about the Green Bank meeting. Drake's account is in chapters 2–3 of Drake and Sobel, *Is Anyone Out There?* and also in his interview in Swift, *SETI Pioneers,* 60–68. Another eyewitness account is in Pearson, "Extraterrestrial Intelligent Life and Interstellar Communication." Walter Sullivan, science editor for the *New York Times,* extensively covered the meeting in his *We Are Not Alone.* Modern overviews are in Morton, "A Mirror in the Sky," and Davies, *The Eerie Silence,* 77–82.

17. David Packard provides a biographical sketch of Oliver in Oliver, *The Selected Papers of Bernard M. Oliver,* ix–xi.

18. *Proposed Studies,* U.S. House of Representatives, April 18, 1961, 1, 214–16, 225–26; "Mankind Is Warned to Prepare for Discovery of Life in Space," *New York Times,* December 15, 1960.

19. George Dugan, "Other-World Bid to Earth Doubted," *New York Times,* June 24, 1960.

20. Davies, *Are We Alone?* 21.

21. Creating life's building blocks is the easy part. Science has yet to understand the origins of systems that process information and replicate themselves. See Davies, *The Eerie Silence*, 29–33. A bigger question than life's origins is the very definition of the term *life*. According to one estimate, *life* has over 300 definitions, each competing for attention, and each can be found wanting. See Kaufman, *First Contact*, 35–39.

22. Lilly and Jeffrey, *John Lilly, So Far,* 102–19.

23. Ibid., 120–21. In the following years, Lilly did not live up to these expectations. The promised interspecies communications never took place. Lilly did report that his dolphins were oversexed, one in particular being very gentle with Lilly's female assistant, allowing himself to be masturbated. Lilly believed that the dolphin's mildness, sexuality, and interest in the assistant pointed to a similarity with humans. Another explanation is possible. When masturbated, the dolphin was merely reacting to stimuli—like a cat purring when petted.

24. E. H. Carr writes: "The essence of man as a rational being is that he develops his potential capacities by accumulating the experience of past generations. Modern man is said to have no larger a brain, and no greater innate capacity of thought, than his ancestors 5000 years ago. But the effectiveness of his thinking has been multiplied many times by learning and incorporating in his experience the experience of the intervening generations" (qtd. in Tosh, *Historians on History*, 56–57).

25. Michaud, *Contact with Alien Civilizations*, 34–35.

26. Herzing, "SETI Meets a Social Intelligence."

27. Wilkinson, *Science, Religion*, 77.

28. Sagan, *Cosmos*, 335–36; Wilkinson, *Science, Religion*, 24, 88.

29. Davies, *The Eerie Silence*, 72–76; Short, *The Gospel from Outer Space*, 3. Instead of seeing religion and science as being at odds, John Hedley Brooke offers a complexity model of nuance in which religion and science influence each other ("Science, Religion, and Historical Complexity"). Ronald Numbers's anthology contains twenty-five essays debunking modern myths of the religion-science nexus. Three of the essays are of note here. David Lindberg's piece challenges the accusation that Christianity caused the demise of ancient science. Michael Shank continues in this vein by insisting that the medieval church did not oppose the growth of science. On the other hand, according to Noah Efron, to claim that Christianity is responsible for modern science is overly simplistic.

30. In 1967, Drake speculated that our galaxy contained 10,000 detectable civilizations. If they were evenly distributed, the closest should lie within 1,000 light-years. See Drake, "Intelligent Life in Other Parts of the Universe," 312.

31. Grinspoon, *Lonely Planets*, 294.

32. Walter Sullivan, "Contact with Worlds in Space Explored by Leading Scientists," *New York Times*, February 4, 1962.

33. Struve, "Astronomers in Turmoil," 22.

34. John W. Finney, "Scientists and Congress Ponder If Life Exists in Other Worlds," *New York Times*, March 23, 1962.

Chapter 5. The 1960s: The Selling of SETI

1. In 1966, Carl Sagan predicted that life would be created within a decade. See Basalla, *Civilized Life in the Universe*, 131.

2. Bracewell, "Life in the Galaxy," 236.

3. Peter Bart, "Symposium on Coast Assays Life on Other Planets," *New York Times*, May 29, 1966.

4. Dickson, *Sputnik*, 118–21, 154–64, 170–77.

5. Paul Gilster, "SETI: The Michaud-Cooper Dialogue," April 20, 2011, http://www.centauri-dreams.org/?p=17625.

6. Grinspoon, *Lonely Planets*, 51–52; Drake, *Intelligent Life in Space*, 72; Morrison, "Interstellar Communication," 124; Calvin, "Chemical Evolution," 73.

7. W. Sullivan, *We Are Not Alone*, 280.

8. Stanton T. Friedman, Prepared Statement to the U.S. House of Representatives, Committee on Science and Astronautics, July 29, 1968, files.ncas.org/ufosymposium/friedman.html. Friedman is a leading ufologist.

9. Morrison, "Interstellar Communication," 122.

10. Oliver and Billingham, *Project Cyclops*, iii.

11. Morrison, "Interstellar Communication," 127–28.

12. Von Hoerner, "General Limits of Space Travel," 204.

13. Cameron, *Interstellar Communication*, 1.

14. W. Sullivan, *We Are Not Alone*, 288–89.

15. Ibid., 291.

16. According to Lutheran theologian Ted Peters, the belief in extraterrestrials' evolutionary progress is "the ETI Myth" of our time. He sees the "Myth" as "a displaced human desire for a comprehensive yet friendly cosmos." See his "Myth in the Heart of Science," April 13–14, 2012, www.pcts.org/meetings/2012/PCTS2012Apr-Peters-Myth-Sci.pdf.

17. Bracewell, "Communication from Superior Galactic Communities," 244. In 1974, Bracewell published *The Galactic Club*, in which he claimed that humanity would receive the galactic culture and asked whether this would "set us to a new and higher path."

18. Hoyle, "Can We Learn from Other Planets?"

19. Michaud writes in *Contact with Alien Civilizations*: "The Galactic Club . . . is an idealized vision of how international relations should be, rather than how our historical experience tells us they are. It is only one of many possible models of relationships among technological civilizations. Others include isolation, anarchy, centerless cooperation, alliance, federation, dominance, and empire" (316).

20. Shklovskii and Sagan, *Intelligent Life in the Universe*, 413.

21. Guthke, *The Last Frontier*, 346–47; Basalla, *Civilized Life in the Universe*, 66.

22. Dick, *Life on Other Worlds*, 45; Hoyt, *Lowell and Mars*, 214–17.

23. Altschuler, *Children of the Stars*, 140–41.

24. Wilkinson writes: "the emergence of intelligence on Earth was dependent of such things as the onset of photosynthesis, the emergence of cells, the growth of

multicellularity, the arrival of sex, and the invasion of the land at the most basic level. This is not even to mention things such as the development of a nervous system and other essential organs" (*Science, Religion*, 76).

25. Simpson, "The Nonprevalence of Humanoids," 214.

26. Drake, "Methods of Communication," 120.

27. Plank, *Emotional Significance of Imaginary Beings*, 101.

28. Morrison, "Interstellar Communication," 124; Drake, *Intelligent Life in Space*, 72.

29. Bergreen, *Voyage to Mars*, 102–3.

30. Oliver and Billingham, *Project Cyclops*, 13–14; Morrison, Billingham, and Wolfe, "The Search for Extraterrestrial Intelligence," 13; Trudy E. Bell, "The Grand Analogy," *Cosmic Search* 2, no. 1 (Winter 1980), http://www.bigear.org/CSMO/HTML/CS05/cs05p02.htm.

31. Shklovskii and Sagan, *Intelligent Life in the Universe*, 357.

32. Drake, *Intelligent Life in Space*, 58, 71–72.

33. Davidson, *Carl Sagan*, 29–30; Sagan, *The Dragons of Eden*, 242–43.

34. Davidson, *Carl Sagan*, 238–40.

35. Bracewell, "Life in the Galaxy," 240–42; Bracewell, *The Galactic Club*, 69–83.

36. Morrison, "Interstellar Communication," 128–29.

37. Drake and Sobel, *Is Anyone Out There?* 131–32.

38. Bracewell interview in Swift, *SETI Pioneers*, 149–51.

39. Sagan has been the subject of three biographies: Davidson's, Poundstone's, and Spangenburg and Moser's.

40. Sagan, "Direct Contact among Galactic Civilizations."

41. Ibid., 209–10; Davidson, *Carl Sagan*, 131–32.

42. Carl Sagan folder, box 10, Donald H. Menzel Papers, American Philosophical Society, Philadelphia.

43. McDonough, *The Search for Extraterrestrial Intelligence*, 49; Davidson, *Carl Sagan*, 178.

44. See note 42.

45. Appleyard, *Aliens*, 104; Poundstone, *Carl Sagan*, 128–32.

46. Davidson, *Carl Sagan*, 196–99.

Chapter 6. Soviet SETI: ET Is a Communist

1. In 1964 the Soviet Ministry of Defense exuded Soviet headiness when it forecast a Russian moon landing in 1968–70 as well as the creation of a space station with a crew of thirty to fifty in 1972–75, flights to Mars and Venus of manned spacecraft in 1975–80, and a Mars landing in 1980–90. See Krieger, "Space Programs of the Soviet Union," 218.

2. Andrews, *Red Cosmos*, 17–18; Young, *The Russian Cosmists*, 3–4.

3. Rosenthal, *The Occult in Russian and Soviet Culture*, 1–32; Holquist, "Philosophical Bases of Soviet Space Exploration," 2–4; Lytkin, Finney, and Alepko, "Tsiolkovsky, Russian Cosmism, and Extraterrestrial Intelligence," 370.

4. Hagemeister, "Russian Cosmism," 187–89.

5. Holquist, "Philosophical Bases of Soviet Space Exploration," 2–4.

6. Andrews, *Red Cosmos,* 31–66.

7. Ibid., 84–96.

8. Young, *The Russian Cosmists,* 151.

9. W. Sullivan, *We Are Not Alone,* 226.

10. Lytkin, Finney, and Alepko, "Tsiolkovsky, Russian Cosmism, and Extraterrestrial Intelligence," 371–72; Webb, *Where Is Everybody?* 22–23.

11. Webb, *Where Is Everybody?* 22–23.

12. Finney, Finney, and Lytkin, "Tsiolkovsky and Extraterrestrial Intelligence," 747.

13. Conquest, *Stalin,* 211; Hingley, *Joseph Stalin,* 200–201.

14. Easterbrook, "Are We Alone?" 26; Brin, "Mystery of the Great Silence," 139; Bobrovnikoff, "Soviet Attitudes," 456.

15. Shklovskii and Sagan, *Intelligent Life in the Universe,* vii–viii, 437. In his memoir, *Five Billion Vodka Bottles to the Moon,* Shklovskii condemns the politicians and scientists who lack concern for nuclear weapons, calling them "cannibals."

16. Kardashev interview in Swift, *SETI Pioneers,* 178–97.

17. Kardashev, "Transmission of Information by Extraterrestrial Civilizations."

18. The minutes are in Tovmasyan, *Extraterrestrial Civilizations;* Ambartsumyan is quoted on page 3.

19. Kardashev interview in Swift, *SETI Pioneers,* 184.

20. Drake and Sobel, *Is Anyone Out There?* 102–4.

21. Ibid., 107.

22. Troitskii interview in Swift, *SETI Pioneers,* 198–208.

23. Poundstone, *Carl Sagan,* 140–41; D. P. Cruikshank, "Vassili Ivanovich Moroz—An Appreciation," www.lpi.usra.edu/meetings/lpsc2005/pdf/1979.pdf.

24. Basalla, *Civilized Life in the Universe,* 153.

25. Sagan published the proceedings of the conference in *Communication with Extraterrestrial Intelligence;* see also Poundstone, *Carl Sagan,* 140–55.

26. "CETI Questionnaire," *Spaceflight* 15, no. 4 (1973): 137–38.

27. Drake, "On Hands and Knees."

28. Drake and Sobel, *Is Anyone Out There?* 110–16.

29. "The Soviet SETI Programme," *Icarus* 26 (1975): 377–85.

30. The Tallinn Conference took place at a time when Congress had prohibited NASA from supporting SETI. Only grants from the Sloan Foundation and the Planetary Society allowed any of the Americans to attend. McDonough, *The Search for Extraterrestrial Intelligence,* 155–56.

31. Frank Drake, "SETI Tallinn-81," *Cosmic Search* 4, no. 1 (First half 1982), http://www.bigear.org/CSMO/HTML/CS13/cs13p04.htm; Drake and Sobel, *Is Anyone Out There?* 155–57.

32. W. T. Sullivan, "SETI Conference at Tallinn," 350.

33. Young, *The Russian Cosmists,* 219, 224–26.

34. Osiatynski, *Contrasts,* 23.

Chapter 7. The 1970s: Gaining Respect

1. John Billingham, phone interview by the author, September 22, 2005. When I asked Billingham why he joined SETI, he said that he expected great things from contact, but he did not specify what he had in mind. When later asked the same question, he gave the same vague answer. I surmised that being a lifelong bureaucrat, Billingham was careful in his public utterances.

2. Gerrit Verschuur, e-mail to the author, September 29, 2014; Verschuur, "A Search for Narrow Band 21-cm Wavelength Signals."

3. Billingham, interview by the author. NASA historian Steven J. Dick interviewed Billingham on September 17, 1992, and June 8, 1993; Freedom of Information Act [hereafter cited as FOIA], NASA. Billingham also told his story in Swift, *SETI Pioneers*, 246–78.

4. Oliver and Billingham, *Project Cyclops*, 1971. Oliver lectured on Project Cyclops at the 1971 Byurakan Conference; see Sagan, *Communication with Extraterrestrial Intelligence*, 279–302. Oliver also summarized Project Cyclops in "Technical Considerations in Interstellar Communication." Bracewell provides a good summary in *The Galactic Club*, 41–47.

5. Rood and Trefil, *Are We Alone?* 154.

6. According to George Basalla, Project Cyclops suffered from the fallacious notion that technology was moving "progressively toward goals predetermined by the universal laws of science" (*Civilized Life in the Universe*, 187). To believe in the convergence of technologies regardless of the difference of life-forms ignores the influence of culture.

7. Appleyard, *Aliens*, 235.

8. "Seven Ways to Compute the Relative Value of a U.S. Dollar Amount—1774 to Present," http://www.measuringworth.com/uscompare.

9. Greenstein, *Astronomy and Astrophysics* 49–52, 122–23.

10. Fletcher, "The Space Program," 45–47.

11. FOIA, NASA: Fletcher to Hans Mark, December 2, 1972.

12. FOIA, NASA: Meeting Record of "Proposal to Undertake Search for Extraterrestrial intelligence," September 11, 1973.

13. FOIA, NASA: Oliver to Fletcher, September 20, 1973, and Fletcher to Oliver, October 4, 1973.

14. FOIA, NASA: Oliver to Fletcher, April 16, 1974.

15. On page 3 of the Cyclops Report, Oliver wrote a succinct justification of Project Cyclops and, in fact, of SETI itself: "To justify such an effort, which may require billions of dollars and decades of time, we must truly believe that other intelligent life exists and that contact with it would be enormously stimulating and beneficial to mankind."

16. FOIA, NASA: Mark to Fletcher, April 8, 1974.

17. Phillip Morrison interview in Swift, *SETI Pioneers*, 43.

18. For the Boston SETI Symposium, I have relied on Berendzen, *Life beyond Earth*.

19. Drake, "On Hands and Knees," 27–29.

20. Morrison interview in Swift, *SETI Pioneers*, 42.

21. Drake and Sobel, *Is Anyone Out There?* 179–80; Berendzen, *Life beyond Earth*, 85.

22. Basalla, *Civilized Life in the Universe*, 113–14. The writer was Eric Burgess.

23. Spangenburg and Moser, *Carl Sagan*, 110.

24. Sagan, Sagan, and Drake, "A Message from Earth." David Grinspoon doubts that extraterrestrials will understand the Pioneer plaque. He writes: "It is no coincidence that many SETI pioneers have been members of skeptics organizations advocating a hard-core materialist world view, not tolerant of alternative interpretations of nature. A reductionist, universalist scientific philosophy is an unspoken assumption of SETI." In fact, much of SETI is "hugely dependent on our own biases, assumptions, and culture" ("SETI and the Science Wars," 56). Regarding convergent evolution, Giancarlo Genta notes that although sharks, ichthyosaurs, and dolphins converge in shape, "fish, reptiles, and mammalians all belong to the chordate phylum and are genetically very closely related" (*Lonely Minds*, 195). They have the genes of a common ancestor. Genta asks whether beings with nothing in common except life can converge.

25. Sagan, *Cosmos*, 331–32.

26. Davidson, *Carl Sagan*, 342–43; Smith, *Possibility of Intelligent Life*, 76–77.

27. Sagan, *The Cosmic Connection*, 20–27; Drake and Sobel, *Is Anyone Out There?* 175–79; Poundstone, *Carl Sagan*, 134–35.

28. An acid assessment of Sagan appeared in the *Atlantic Monthly* in which the mathematician Alfred Adler remarked on Sagan's alleged ego. According to Adler, the plaque's message conflated astronomy with social and political activism. Adler dismissed Sagan as "immodest," a self-styled "Renaissance Man" who furthermore believed himself "a master of politics and economics." The real Sagan, in Adler's view, was merely "a gifted, highly trained, opportunistic, humorless, and unimaginative ass." See Adler, "Behold the Stars," 226.

29. Drake and Sobel, *Is Anyone Out There?* 180–83.

30. Dick, *Life on Other Worlds*, 243.

31. Drake and Sobel, *Is Anyone Out There?* 184–85; Davidson, *Carl Sagan*, 274.

32. Sagan et al., *Murmurs of Earth*, preface.

33. Ibid.

34. Richard Gray, "Voyager Spacecraft: What Will It Teach the Universe about Mankind?" *The Telegraph*, September 13, 2013, http://www.telegraph.co.uk/science/space/10307179/Voyager-spacecraft-what-will-it-teach-the-universe-about-mankind.html.

35. Eberhart, "Voyager's Message," 211, 221; Spangenburg and Moser, *Carl Sagan*, 131.

36. Kraus interview in Swift, *SETI Pioneers*, 239–42.

37. Gray, *The Elusive Wow*, xv–xvi, 6–9.

38. Ehrman, "'WOW!'—A Tantalizing Candidate," 53–63.

39. Billingham interview by Dick, September 12, 1990.

40. For the workshops, see Morrison, Billingham, and Wolfe, *The Search for Extraterrestrial Intelligence* and "The Search for Extraterrestrial Intelligence."

41. Drake and Sobel, *Is Anyone Out There?* 140.

Chapter 8. SETI in NASA: Rise and Fall

1. David Viewing's "Directly Interacting Extraterrestrial Communities" makes roughly the same argument as Hart, but it received less attention.

2. Webb, *Where Is Everybody?* 17–25. Webb concentrates on possible solutions to the Fermi Paradox. He places the fifty he considers into three broad categories: "They Are Here," "They Have Not Yet Communicated," and "They Do Not Exist."

3. Papagiannis, "The Fermi Paradox," 437.

4. Tipler, "Additional Remarks on Extraterrestrial Intelligence," 290. Tipler has also cried censorship, claiming that when he submitted anti-SETI articles to scholarly journals, the referees included Sagan and Morrison and their negative responses prevented publication.

5. The proceedings of the Montreal sessions are covered in Papagiannis, *The Search for Extraterrestrial Life*, and also in Papagiannis, "Strategies for the Search for Life: A Joint Session of the International Astronomical Union," *Cosmic Search* 2, no. 1 (Winter 1980), http://www.bigear.org/CSMO/HTML/CS05/cs05p24.htm.

6. According to Ian A. Crawford, the problem of interstellar travel already has technological solutions, such as anti-matter rockets. Crawford adds that surprises mark the history of technology and that this has profound implications for the SETI debate. See Crawford, "Interstellar Travel," 67.

7. White, *The SETI Factor*, 70–71.

8. Hart is now a white separatist. Although his political incorrectness is irrelevant to SETI, it does show that he has continued to be a contrarian.

9. For Drake, see Michaud, *Contact with Alien Civilizations*, 43; for Sagan, see Mc-Donough, *The Search for Extraterrestrial Intelligence,* 202; Wolfe, "On the Question of Interstellar Travel," 452; for Gould, see Webb, *Where Is Everybody?* 24.

10. Another anti-SETI blast came from Robert T. Rood (1942–2011) and James S. Trefil (b. 1938) in their 1981 book *Are We Alone?* Rood and Trefil revisited the Drake Equation, insisting it needed additional variables. By increasing the variables, each of which might fail, they reduced N, the number of extraterrestrial civilizations.

11. Frosch, Introduction, xiii–xv.

12. Papagiannis, "Historical Introduction," 9; Papagiannis, "The Hunt Is On," 210.

13. Guillermo Lemarchand, "A New Era in the Search for Life in the Universe," Proceedings of the Bioastronomy Conference of the IAU Commission 51, held in Hawaii, August 1999, http://www.iar-conicet.gov.ar/SETI/bio99-2.pdf; Brin, "Mystery of the Great Silence," 148–51.

14. National Research Council, Astronomy Survey Committee, *Astronomy and Astrophysics for the 1980's*, 17–19, 28, 90–91, 150.

15. Brin, "Mystery of the Great Silence," 152–59.

16. Drake and Sobel, *Is Anyone Out There?* 140.

17. Richard Severo, "William Proxmire, Maverick Democratic Senator from Wisconsin, Is Dead at 90," *New York Times*, December 16, 2005, http://www.nytimes.com/2005/12/16/national/16proxmire.html?pagewanted=all&_r=1&.

18. Drake and Sobel, *Is Anyone Out There?* 191–92.

19. The proceedings are in *Extraterrestrial Intelligence Research*, Hearings before the House Committee on Science and Technology, September 19–20, 1978.

20. Billingham interview by Dick, September 12, 1990.

21. U.S. Congress, Senate, *Congressional Record*, July 30, 1981.

22. Dick and Strick, *The Living Universe*, 143.

23. Billingham interview by Dick, June 1, 1992; Drake and Sobel, *Is Anyone Out There?* 191–96.

24. Marcia Smith, "Confronting Political Realities: The Federal Funding Process: A SETI Case History," *Cosmic Search* 2, no. 3 (Summer 1980), http://www.bigear.org/CSMO/HTML/CS07/cs07p30.htm.

25. Billingham interviews by Dick, September 12, 1990, and June 1, 1992; Gray, *The Elusive Wow*, 136–37; Dick and Strick, *The Living Universe*, 144–45.

26. Kerr, "NASA's Search for ETs Hits a Snag on Earth," 249–50.

27. Howard Blum, "SETI, Phone Home," *New York Times*, October 21, 1990, 35.

28. William J. Broad, "Hunt for Aliens in Space: The Next Generation," *New York Times*, February 6, 1990.

29. Dick, "The Search for Extraterrestrial Intelligence," 119–29; Papagiannis, "The Hunt Is On," 216–18; Basalla, *Civilized Life in the Universe,* 168–69.

30. Michaud, "A Unique Moment"; Genta, *Lonely Minds,* 279–82; Koerner and LeVay, *Here Be Dragons*, 169.

31. Ernst Mayr, letter to *Science*, March 12, 1993, 1522–23. According to Davies, evolutionary "purists" are leery of directionality, seeing it as a means of slipping design into evolution (*Are We Alone?* 76).

32. Frank Drake, letter to *Science*, April 23, 1992, 424–25; Frank Drake, "Space Travel and Life beyond the Earth," *Cosmic Search* 3, no. 2 (April, May, June 1981), http://www.bigear.org/CSMO/HTML/CS10/cs10p05.htm; see also Russell, "Speculations on the Evolution of Intelligence."

33. Verschuur, "If We Are Alone"; White, *The SETI Factor*, 31; Drake and Sobel, *Is Anyone Out There?* 27.

34. Shostak, *Sharing the Universe*, 160.

35. U.S. Congress, Senate, *Congressional Record*, September 22–23, 1993. In October 2005 the author sent an e-mail to Senator Bryan, asking whether skepticism toward SETI played a role in his actions. Senator Bryan replied that concern with the federal deficit was his only motive.

36. "Space Agency Is Faulted on Contractor Fraud," *New York Times*, October 7, 1993.

37. Unfortunately for SETI, its leading advocates have not hidden their elitist feelings, often making wisecracks about the lack of intelligence on Earth. Television has especially drawn scorn. Witness the following from Sagan: "The reception of a housewife's daytime television serial, from interplanetary space, would undoubtedly interest terrestrial radio astronomers." Shklovskii and Sagan, *Intelligent Life in the Universe,* 435.

38. Report from Gerrit Verschuur to author, May 19, 2014.

39. Garber, "Searching for Good Science," 5.

40. FOIA, NASA, Richard Bryan to Daniel Golden, NASA Administrator, November 10, 1993.

41. Oliver, "Reflections," 70.

42. Dick, "The Search for Extraterrestrial Intelligence," 130.

43. Bowyer, "SERENDIP," 99–104; Genta, *Lonely Minds*, 275.

44. McDonough, *The Search for Extraterrestrial Intelligence*, 148–49.

45. Ibid., 146–47.

46. Leigh and Horowitz, "Millions and Billions of Channels," 119–23; Genta, *Lonely Minds*, 276.

47. Hay River, Northwest Territories: SETI Website, http://www.tranquileye.com/truth/casebook/hay_river_seti.html.

48. Shostak, *Confessions of an Alien Hunter*, 146–47; Pierson, "Birth of the SETI Institute," 4–6.

49. Shostak, *Confessions of an Alien Hunter*, 155–57.

Chapter 9. After NASA: Revival, Distress, and Hope

1. Billings, *Five Billion Years of Solitude*, 9–10.

2. Wattenberg, *Fewer*, 5–16; Barbara Crossette, "Population Estimates Fall as Poor Women Assert Control," *New York Times*, March 10, 2002, http//www.nytimes.com/2002/03/10/international/10POPU.html?ex=1016732845&ei=1&en+9c84417e9c2b44a3.

3. Easterbrook's "We're All Gonna Die" provides a laundry list of calamities ranging from voracious black holes to killer viruses.

4. See Rees, *Our Final Hour*; also Darling and Schulze-Makuch, *Megacatastrophes*.

5. Grinspoon provides a good summary and critique of the Rare Earth thesis (*Lonely Planet*, 143–48). Jill Tarter's retort to Ward and Brownlee is that they argue from a single example ("Exoplanets, Extremophiles, and the Search for Extraterrestrial Intelligence," 8).

6. Pitta, "The Forbes Four Hundred."

7. Alan M. MacRobert, "SETI Searches Today," *Sky and Telescope*, September 24, 2011, http://www.skyandtelescope.com/resources/seti/3304561.html?showAll=y&c=y; David Schrieberg, "Hollywood Gives SETI Its Big Break," *Christian Science Monitor*, August 6, 1997, http://www.csmonitor.com/1997/0806/080697.us.us.4.html.

8. Shuch, "Project Argus," 201–20; Shuch, "A Half-Century of SETI Science," 7; SETI League, 2014 Annual Report, www.setileague.org/admin/report14.pdf.

9. SETI League, *SearchLites* 19, no.3 (Summer 2013), http://setileague.org/srchlite/vol19no3.pdf. See 2014 Annual Report in note 8.

10. Korpela, "Distributed Processing of SETI Data," 183–86.

11. Rheingold, "You Got the Power."

12. Robert Sanders, University of California, Berkeley, "Searching for ET from Home—UC Berkeley Launches Project to Draw Public into the Search for Extraterrestrial Intelligence," November 25, 1998, http://berkeley.edu/news/media/releases/98legacy/11-25-1998.html.

13. "The Big Crunch: Interview with Dan Werthimer," *Astrobiology Magazine*, December 8, 2003, http://www.astrobio.net/interview/710/the-big-crunch-interview-with-dan-werthimer; "SETI@home Celebrates 10th Anniversary," May 19, 2009, http://www.spaceref.com/news/viewpr.html?pid=28232.

14. J. D. Biersdorfer, "The Search for E.T. Yields Earthly Cheats," *New York Times*, May 24, 2001, http://www.nytimes.com/2001/05/24/technology/the-search-for-et-yields-earthly-cheats.html.

15. MacRobert, "SETI Searches Today." (See note 7 on p. 203.)

16. "SETI Survey Focuses on Kepler's Top Earth-Like Planets," May 16, 2011, http://www.spaceref.com/news/viewpr.html?pid=33572.

17. Alexander and Anderson, "Searching for E.T."

18. Daniel Terdiman, "SETI's Large-Scale Telescope Scans the Skies," *CNET*, December 12, 2008, http://www.cnet.com/news/setis-large-scale-telescope-scans-the-skies.

19. Paul Brest, "A Decade of Outcome-Oriented Philanthropy," *Stanford Social Innovation Review*, Spring 2012, http://www.ssireview.org/articles/entry/a_decade_of_outcome_oriented_philanthropy.

20. M. Mitchell Waldrop, "The Search for Alien Intelligence," *Nature*, July 27, 2011, http://www.nature.com/news/2011/110727/full/475442a.html.

21. Alan M. MacRobert, "The Allen Telescope Array: SETI's Next Big Step," *Sky and Telescope*, November 18, 2011, http://www.skyandtelescope.com/resources/seti/3304581.html.

22. Amrutha Gayathri, "Jodie Foster Joins Alien Hunt," *International Business Times*, August 17, 2011, http://www.ibtimes.com/jodie-foster-joins-alien-hunt-834081; "Jill Tarter, SETI Astronomer, Retiring after 35-Year Alien Hunt," *Huffington Post*, May 22, 2012, http://www.huffingtonpost.com/2012/05/22/jill-tarter-seti-astronomer-retiring-alien-hunt_n_1536566.html (see note 22 above).

23. Leonard David, "SETI Telescope to Help US Air Force Track Space Junk," April 30, 2012, http://www.space.com/15479-seti-telescope-space-junk-search.html.

24. Keith Cowing, "SETI Institute Receives $3.5 Million Donation," November 15, 2012, http://spaceref.com/astrobiology/seti-institute-receives-35-million-donation.html.

25. "Jill Tarter, SETI Astronomer, Retiring after 35-Year Alien Hunt."

26. *Life in the Universe*, Hearing before the Subcommittee on Space and Aeronautics, Committee on Science, U.S. House of Representatives, 107th Congress, 1st sess., July 12, 2001, Serial No. 107-17, http://commdocs.house.gov/committees/science/hsy73839.000/hsy73839_0f.htm.

27. "Testimony of Dr. Steven J. Dick, Hearing on Astrobiology," *Astrobiology Web*, December 4, 2013, http://astrobiology.com/2013/12/testimony-of-dr-steven-j-dick-hearing-on-astrobiology.html.

28. Oliver, "NASA's Role in Man's Future," 5.

29. Lisa Grossman, "Arecibo Radio Telescope May Lose Funding," *SpaceDaily*, November 10, 2006, http://www.spacedaily.com/reports/Arecibo_Radio_Telescope_May_Lose_Funding_999.html.

30. Yudhijit Bhattacharjee, "New Consortium to Run Arecibo Observatory," *Science Insider*, May 20, 2011, http://news.sciencemag.org/funding/2011/05/new-consortium -run-arecibo-observatory.

31. Casey Dreier, "Action Alert," March 24, 2014, http://support.planetary.org/site/ MessageViewer?dlv_id=10981&em_id=6281.0.

32. Dreier, "Breaking Sisyphus' Curse," 22–23.

33. Dan Geraci, "Annual Report to Our Members," June 2012 and June 2013. In its annual report for the 2013 fiscal year, the Planetary Society bragged of an uptick in membership, reversing the downward trend of over a decade.

34. Ekers et al., *SETI 2020*, 232.

35. Wilson Da Silva, "Only a Matter of Time, Says Frank Drake," *Cosmos*, April 7, 2010, http://www.cosmosmagazine.com/planets-galaxies/qa-with-frank-drake.

36. Seth Shostak, "E.T., Phone the Bay Area," *Huffington Post*, July 12, 2011, http:// www.huffingtonpost.com/seth-shostak/et-phone-the-bay-area_b_886765.html.

37. Shostak, *Confessions of an Alien Hunter*, 295.

38. Bob Yirka, "Astrophysicists Apply New Logic to Downplay the Probability of Extraterrestrial Life," July 27, 2011, http://phys.org/news/2011-07-astrophysicists-logic- downplay-probability-extraterrestrial.html.

39. Webb, "Pondering the Fermi Paradox," 306. In April 2015, Penn State University researcher Roger Griffith announced having used NASA's WISE orbiting observatory to search 100,000 galaxies for Dyson spheres. He reported no signs of radiated heat that would indicate a Type III civilization. But he did find fifty galaxies with high levels of heat radiation. Whether the heat is natural or the result of advanced civilizations is unknown. See "Search for Advanced Civilizations beyond Earth Finds Nothing Obvious in 100,000 Galaxies," *Science Daily*, April 14, 2015, www.sciencedaily.com/ releases/2015/04/150414101000.htm.

40. Davies, *The Eerie Silence*, 8. Gerrit Verschuur posits a somewhat similar scenario in which advanced extraterrestrials have a consciousness whose superiority renders humans unable to understand. Contact will take place only if extraterrestrials "dumb" down to our level. Verschuur, *Is Anyone Out There?* 339–41.

41. Dick, "Bringing Culture to Cosmos," 467–80.

42. Lisa Grossman, "Astronomers Suggest Crowdsourcing Letters to Aliens," *Wired Science*, February 9, 2011, http://www.wired.com/2011/02/crowdsourced-seti/.

43. Zaitsev, "METI," 400–405.

44. Alan Boyle, "Should We Be Phoning E.T.?" *Cosmic Log*, July 14, 2008, http://re- search.lifeboat.com/cosmic.log.htm; "Doritos Makes History with World's First Extra Terrestrial Advertisement," *Science Daily*, June 12, 2008, http://www.sciencedaily.com/ releases/2008/06/080612122817.htm.

45. "Humanity Responds to 'Alien' Wow Signal, 35 Years Later," August 17, 2012, http://www.space.com/17151-alien-wow-signal-response.html.

46. Zaitsev, "METI," 399.

47. Vakoch, "Integrating Active and Passive SETI Programs," 263–64.

48. David Brin, "Shouting at the Cosmos," September 2006, https://lifeboat.com/ ex/shouting.at.the.cosmos.

49. Kaufman, *First Contact,* 177.

50. Zaitsev, "METI," 423–24.

51. Kaufman, *First Contact,* 177; Brin, "A Contrarian Perspective on Altruism," 447.

52. Denning, "Unpacking the Great Transmission Debate," 245.

53. Filling in the terms of the Drake Equation is a waste of time, since some of the values are mostly educated guesses. Any reasonable estimate for N can pass muster.

54. Drake and Sobel, *Is Anyone Out There?* 170–73.

55. Ekers et al., *SETI 2020,* 292.

56. Betts, "Looking for ET Using Laser Light," 8.

57. Betts, "We Make It Happen!"; Ross and Kingsley, "Optical SETI," 149, 152.

58. See note 3.

59. Betts, "Looking for ET Using Laser Light," 8; MacRobert, "SETI Searches Today" (see note 7 on p. 203). Another OSETI convert is Geoffrey Marcy (b. 1954) of UC Berkeley, who is better known for his discovery of exoplanets, planets outside of the solar system. Marcy's quest for exoplanets has produced data that can also be checked for continuous laser waves. "The Search for Extraterrestrial Intelligence at Berkeley, *SEVENDIP,*" http://seti.berkeley.edu/opticalseti.

60. Verschuur, *Is Anyone Out There?* 311–14.

61. J. Tarter, "Exoplanets, Extremophiles, and the Search for Extraterrestrial Intelligence," 5–8.

62. Appleyard, *Aliens,* 223–24. The existence of extremophiles weakens Rood and Trefil's anti-SETI polemic, *Are We Alone?* In explaining the origin of terrestrial life, Rood and Trefil note the concentration of organic nutrients in tidal pools. It is now surmised that life could have emerged in the ocean depths, using heat from the Earth's core instead of the sun. The same may be true for other worlds. See Genta, *Lonely Minds,* 71. Corey S. Powell in the September 2015 issue of *Popular Science* is euphoric over Europa, which he deems "the wettest known world in the solar system" and the best bet in the solar system for non-terrestrial life ("Europa or Bust: Searching for Life in Jupiter's Orbit," September 21, 2015, http://www.popsci.com/europa-or-bust).

63. Bhattacharjee, "Mr. Borucki's Lonely Road to the Light."

64. "A Roadmap for Planet-Hunting," *The Economist,* April 8, 2000; J. Tarter, "Exoplanets, Extremophiles, and the Search for Extraterrestrial Intelligence," 4–5.

65. Ken Croswell, "Gentle or Jumping? The Varied Lives of Hot Jupiters," *Scientific American,* April 1, 2013, http://www.scientificamerican.com/article/the-varied-lives -of-hot-jupiters.

66. "Kepler and K2," July 2015, http://www.nasa.gov/kepler.

67. Robert Sanders, "Astronomers Answer Key Question: How Common Are Habitable Planets?" *UC Berkeley News Center,* November 4, 2013, http://newscenter.berkeley. edu/2013/11/04/astronomers-answer-key-question-how-common-are-habitable-planets. The authors of the study are Geoffrey Marcy, University of Hawaii astronomy professor Andrew Howard, and UC Berkeley graduate student Erik Petigura. They define "earth-size" planets as being one to two times the size of the Earth with an habitable zone receiving as much as four times the star light of the Earth and as little as one quarter.

68. Billings, *Five Billion Years of Solitude,* 222–23. Ben Zuckerman argues that if NASA can construct a Terrestrial Planet Finder, so can advanced aliens, assuming they exist. See his "Why SETI Will Fail," *Mercury Magazine,* September–October 2002, http://www.astrosociety.org/pubs/mercury/31_05/contents.html.

69. J. D. Harrington and Michele Johnson, "NASA Ends Attempts to Fully Recover Kepler Spacecraft, Potential New Missions Considered," August 15, 2013, www.seti.org/seti-institute/news/nasa-ends-attempts-fully-recover-kepler-spacecraft-potential-new-missions; Michael Lemonick, "The Kepler Space Telescope May Be Dead, but Its Planet-Hunting Mission Continues," *TIME: Science and Space,* August 16, 2013, http://science.time.com/2013/08/16/the-kepler-space-telescope-may-be-dead-but-its-planet-hunting-mission-continues; Mike Wall, "NASA's Exoplanet-Hunting Kepler Space Telescope Gets New Mission," May 16, 2014, www.space.com/25913-nasa-kepler-telescope-new-mission.html.

70. "How Engineers Modified the Spitzer Telescope to Probe Exoplanets," September 24, 2013, http://spaceref.com/exoplanets/how-engineers-modified-the-spitzer-telescope-to-probe-exoplanets.html.

71. "Our Scientists," http://www.seti.org/seti-institute/about-seti/scientists?page=6.

72. Davies, "Life, Mind, and Culture," 386.

73. Shostak, "Astrobiology Growing Pains," 14–15.

74. DeVore, "Making Science Education Exciting," 15–16; DeVore et al., "Educating the Next Generation of SETI Scientists."

75. Shostak, "Tuning In to Science," 15–16.

76. "Science and Technology Celebrated at SETIcon II," June 25, 2012, http://www.seti.org/seti-institute/news/science-and-technology-celebrated-seticon-ii-three-day-festival-ends-high-note; Bob Yirka, "Alien Life Searchers Conference SETICon 2 Held in Santa Clara," June 25, 2012, http://phys.org/news/2012-06-alien-life-searchers-conference-seticon.html.

77. "Join the SETI Search," February 2009, http://www.ted.com/talks/jill_tarter_s_call_to_join_the_seti_search/transcript; "Jill Tarter Wins TED Award," *SETI Institute Explorer* 6, no. 1 (2009): 18–19.

78. "Join the Quest," *SETIQuest,* March 23, 2010, http://setiquest.org/join-the-quest; Gerry Harp, "Life at the SETI Institute: DIY—Do Your Own SETI Searches with Seti-Quest Data and Software," *Huffington Post,* September 28, 2011, http://www.huffingtonpost.com/seti-institute/life-at-the-seti-institut_b_986273.html.

79. "Discussion Forums," *SETIQuest,* February 9, 2012–March 4, 2012, and June 26, 2012, http://setiquest.org/forum/topic/farewell#comment-3115.

80. "The Wow Factor," *Economist,* March 10, 2012, 92; Michael Crider, "Alien Hunting Goes Crowd-Sourced with SETILive," February 29, 2012, http://www.slashgear.com/alien-hunting-goes-crowd-sourced-with-setilive-29216354.

81. Seth Shostak, e-mail to author, August 13, 2015.

82. John Wenz, "What Alien Hunters Plan to Buy with $100 Million," July 20, 2015, http://motherboard.vice.com/read/what-alien-hunters-can-buy-with-100-million.

83. Zeeya Merali, "Search for Extraterrestrial Intelligence Gets a $100-Million Boost," July 20, 2015, http://www.nature.com/news/search-for-extraterrestrial-intelligence-gets-a-100-million-boost-1.18016.

84. Dennis Overbye, "Stephen Hawking Joins Russian Entrepreneur's Search for Alien Life," *New York Times*, July 20, 2015.

85. Douglas MacMillan and Gautam Naik, "Yuri Milner to Fund $100 Million Search for Intelligent Alien Life," July 20, 2015, *Wall Street Journal*, July 20, 2015.

86. "Yuri Milner and Stephen Hawking Announce $100 Million Breakthrough Initiative to Dramatically Accelerate Search for Intelligent Life in the Universe," July 20, 2015, http://www.spaceref.com/news/viewpr.html?pid=46384.

87. Laura Geggel, "Is E.T. Calling? Massive Search Will Scour Cosmos for Intelligent Aliens," July 20, 2015, http://www.livescience.com/51607-seti-breakthrough-listen-initiative.html; Max Taves, "Russian Billionaire Patiently Listens for Alien Sounds," October 7, 2015, http://www.cnet.com/news/russian-billionaire-patiently-listens-for-alien-sounds.

88. Robert Sanders, "Internet Investor Yuri Milner Joins with Berkeley in $100 Million Search for Extraterrestrial Intelligence," *Berkeley News*, July 20, 2015, http://news.berkeley.edu/2015/07/20/breakthrough-search-for-extraterrestrial-intelligence; Alan MacRobert, "Breakthrough Listen: Giant Leap for SETI," *Sky & Telescope*, July 20, 2015, http://www.skyandtelescope.com/astronomy-news/breakthrough-listen-a-giant-leap-for-seti; see also note 84 above.

89. "Green Bank Telescope Joins "Breakthrough Listen" to Vastly Accelerate Search for Intelligent Life in the Universe," July 20, 2015, http://public.nrao.edu/news/pressreleases/gbt-breakthrough-listen.

90. See note 83 above.

91. "CSIRO and Internet investor Yuri Milner Strike Deal for ET Search," July 21, 2015, http://www.csiro.au/en/News/News-releases/2015/CSIRO-and-Internet-investor-Yuri-Milner-strike-deal-for-ET-search.

92. See note 87 above.

93. Alan MacRobert, "Breakthrough Listen: Giant Leap for SETI," *Sky & Telescope*, July 20, 2015, http://www.skyandtelescope.com/astronomy-news/breakthrough-listen-a-giant-leap-for-seti.

94. Ibid., and also note 87 above.

95. See notes 83 and 93 above.

96. Jeff Foust, "A Funding Breakthrough for SETI," *The Space Review*, August 17, 2015, http://www.thespacereview.com/article/2807/1.

97. Heather Somerville, "Russian Billionaire Yuri Milner Gives $100M to UC Berkeley to Discover Alien Life," *San Jose Mercury News*, July 20, 2015, http://www.mercurynews.com/business/ci_28512628/russian-billionaire-yuri-milner-gives-100-million-uc; see also note 85 above.

98. Matt Vella, "Yuri Milner: Why I Funded the Largest Search for Alien Intelligence Ever," *Time*, July 20, 2015, http://time.com/3964238/yuri-milner-breakthrough-listen.

Conclusion

1. Although ET may be beaming "coherent neutrinos, gravitational waves, or 'subspace radio,'" these technologies are unknown to the early twenty-first century and irrelevant to SETI. Leigh and Horowitz, "Millions and Billions of Channels," 108–9.

2. Genta, *Lonely Minds*, 178; Ekers et al., *SETI 2020*, 63–64.

3. McDonough, *The Search for Extraterrestrial Intelligence*, 137–47.

4. Most journalists whom Frank White interviewed felt that the SETI Protocols will not hold up long. See White, *The SETI Factor*, 127. When NASA in August 1996 announced finding Allan Hills 84001, a meteorite from Mars that seemed to contain fossilized bacteria, the media excitedly proclaimed the possible discovery of Martian life. To this day, though, there is no scientific consensus on the meteorite. Nonetheless, President Clinton lost no time in publicly hailing the event. One can imagine the political bombast if an extraterrestrial signal were received. See Wilkinson, *Science, Religion*, 9–10.

5. Shostak, *Confessions of an Alien Hunter*, 3.

6. Ibid., 211–12.

7. Davies, *The Eerie Silence*, 183–85.

8. Harrison, "After Contact—Then What?" 505–6.

9. Shostak, *Confessions of an Alien Hunter*, 243.

10. Vakoch, "Reactions to Receipt," 742; Ekers et al., *SETI 2020*, 279–80.

11. The modern Mayan languages gave hints to the meaning of the classical Mayan glyphs. This advantage would be absent with ET. See Finney and Bentley, "A Tale of Two Analogues."

12. D. Tarter, "Can SETI Fulfill the Value Agenda of Cultural Anthropology?" 92–93.

13. Harrison, "After Contact—Then What?" 503. Super-ufologist Stanton Friedman will be leading the "I told you so" brigade. Ted Peters writes that Friedman "already lives in a post-contact world. He's upset with establishment scientists because they rely upon SETI rather than UFO reports for their information regarding alien contact." Ted Peters, "Will ETI Contact Put an End to Our World's Religions?" 2011 MUFON Symposium, http://tedstimelytake.com/wp-content/uploads/2013/03/ETIContactReligions.pdf.

14. Tough, "An Extraordinary Event," 1–6.

15. Ruse, "Is Rape Wrong on Andromeda?"

16. Dick, "Cosmic Evolution," 40; Jill Tarter unloads on organized religion in "SETI and the Religions of Extraterrestrials."

17. Raymo, *Skeptics and True Believers*, 63.

18. Sagan, *The Cosmic Connection*, 153.

19. White, *The SETI Factor*, 133.

20. Harrison, *After Contact*, 297; Davies, "Transformations in Spirituality and Religion," 51–52; Genta, *Lonely Minds*, 24–36.

21. Peters, "Will ETI Contact Put an End to Our World's Religions?" (See note 13 above.)

22. See Bainbridge, "Cultural Beliefs about Extraterrestrials"; Pettinico, "American

Attitudes about Life beyond Earth"; and Consolmagno and Mueller, *Would You Baptize an Extraterrestrial?* 255.

23. "SETIcon 2012: Would Discovering ET Destroy Earth's Religion?" (DVD, SETI Institute, 2012).

24. "Vatican Astronomer Says Believing in Aliens Does Not Contradict Faith in God," press release, May 13, 2008, http://vaticandiplomacy.wordpress.com/2008/05/13/vatican-astronomer-says-believing-in-aliens-does-not-contradict-faith-in-god.

25. O'Meara, *Vast Universe*, 24, 47–48.

26. Shostak, *Confessions of an Alien Hunter*, 280.

27. Gerrit Verschuur writes: "We cannot know in advance what the dominant life forms on other planets will be like, what they might be interested in doing, or what they have to tell us. The only certainty is that we know absolutely nothing about them" (*Is Anyone Out There?* 215).

28. Brin, "A Contrarian Perspective on Altruism," 448.

29. David Lasser's brief for space travel can apply to SETI. See his *The Conquest of Space*, 144–53.

BIBLIOGRAPHY

Adler, Alfred. "Behold the Stars." In Goldsmith, *The Quest for Extraterrestrial Life*, 224–27.

Aldiss, Brian W. *Trillion Year Spree: The History of Science Fiction*. New York: Atheneum, 1986.

Alexander, Amir, and Charlene M. Anderson. "Searching for E.T. and the Cure for Cancer." *Planetary Report* 27, no. 3 (May–June 2007): 6–11.

Altschuler, Daniel R. *Children of the Stars: Our Origin, Evolution and Destiny*. Cambridge, U.K.: Cambridge University Press, 2002.

Andrews, James T. *Red Cosmos: K. E. Tsiolkovskii, Grandfather of Soviet Rocketry*. College Station: Texas A&M Press, 2009.

Appleyard, Bryan. *Aliens: Why They Are Here*. London: Simon and Schuster, 2005.

Asimov, Isaac. "Of Life Beyond: Man's Age-Old Speculations." In Christian, *Extra-Terrestrial Intelligence*, 33–52.

Astrophysics in the New Millennium: Panel Reports. Washington, D.C.: National Academy Press, 2001.

Bailery, Cyril. *The Greek Atomists and Epicurus*. New York: Russell and Russell, 1964.

Bainbridge, William Sims. "Cultural Beliefs about Extraterrestrials: A Questionnaire Study." In Vakoch and Harrison, *Civilizations beyond Earth*, 118–40.

Basalla, George. *Civilized Life in the Universe*. New York: Oxford University Press, 2006.

Baxter, Stephen. "SETI in Science Fiction." In Shuch, *Searching for Extraterrestrial Intelligence*, 351–71.

Beck, Lewis White. "Extraterrestrial Intelligent Life." In Regis, *Extraterrestrials*, 3–18.

Berendzen, Richard, ed. *Life beyond Earth and the Mind of Man*. Washington, D.C.: NASA, 1973.

Bergreen, Laurence. *Voyage to Mars*. New York: Riverhead Books, 2000.

Betts, Bruce. "Looking for ET Using Laser Light." *Planetary Report* 30, no. 4 (July–August 2010): 8–11.

———. "We Make It Happen! The Planetary Society Optical SETI Telescope." *Planetary Report* 26, no. 3 (May–June 2006): 4–7.

Bhattacharjee, Yudhijit. "Mr. Borucki's Lonely Road to the Light." *Science,* May 3, 2013, 542–45.

Billingham, John, ed. *Life in the Universe.* Cambridge, Mass.: MIT Press, 1981.

———. "Who Said What: A Summary and Eleven Conclusions." In Tough, *When SETI Succeeds,* 33–39.

Billingham, John, Roger Heyns, David Milne, Stephen Doyle, Michael Klein, John Heilbron, Michael Ashkenazi, Michael Michaud, Julie Lutz, and Seth Shostak, eds. *Social Implications of the Detection of an Extraterrestrial Civilization.* Mountain View, Calif.: SETI Press, 199.

Billingham, John, and Rudolf Pesek, eds. *Communication with Extraterrestrial Intelligence.* Oxford, U.K.: Pergamon Press, 1979.

Billings, Lee. *Five Billion Years of Solitude: The Search for Life among the Stars.* New York: Current, 2013.

Bobrovnikoff, N. T. "Soviet Attitudes Concerning the Existence of Life in Space." In Wukelic, *Handbook of Soviet Space-Science Research,* 453–72.

Bova, Ben, and Byron Preiss, eds. *Are We Alone in the Cosmos? The Search for Alien Contact in the New Millennium.* New York: Ibooks, 1999.

Bowyer, Stuart. "SERENDIP: The Berkeley SETI Program." In Shuch, *Searching for Extraterrestrial Intelligence,* 99–105.

Boyer, Paul. *By the Bomb's Early Light: American Thought and Culture at the Dawn of the Atomic Age.* New York: Pantheon, 1985.

Bracewell, Ronald N. "Communication from Superior Galactic Communities." In Cameron, *Interstellar Communication,* 243–48.

———. *The Galactic Club: Intelligent Life in Outer Space.* San Francisco: W. H. Freeman, 1974.

———. "Life in the Galaxy." In Cameron, *Interstellar Communication,* 232–42.

Brin, David. "A Contrarian Perspective on Altruism: The Dangers of First Contact." In Shuch, *Searching for Extraterrestrial Intelligence,* 429–49.

———. "The Mystery of the Great Silence." In Bova and Preiss, *Are We Alone in the Cosmos?* 137–59.

Brooke, John Hedley. "Science, Religion, and Historical Complexity." *Historically Speaking: The Bulletin of the Historical Society* 8, no. 5 (May–June 2007): 10–13.

Calvin, Melvin. "Chemical Evolution." In Cameron, *Interstellar Communication,* 33–81.

Cameron, A. G. W., ed. *Interstellar Communication.* New York: W. A. Benjamin, 1963.

Cantril, Hadley. *The Invasion from Mars: A Study in the Psychology of Panic.* New York: Harper and Row, 1966.

Christian, James L., ed. *Extra-Terrestrial Intelligence: The First Encounter.* Buffalo, N.Y.: Prometheus Books, 1976.

Clareson, Thomas D. *Understanding Contemporary American Science Fiction: The Formative Period (1926–1970).* Columbia: University of South Carolina Press, 1990.

Clarke, Arthur C. *Childhood's End.* New York: Harcourt, Brace and World, 1953.

———. *The Exploration of Space.* New York: Harper, 1951.

Cocconi, Giuseppe, and Philip Morrison. "Searching for Interstellar Communications." In Goldsmith, *The Quest for Extraterrestrial Life,* 102–4.

Conquest, Robert. *Stalin: Breaker of Nations*. New York: Penguin, 1991.

Consolmagno, Guy, SJ, and Paul Mueller, SJ. *Would You Baptize an Extraterrestrial? . . . and Other Questions from the Astronomers' In-box at the Vatican Observatory*. New York: Image, 2014.

Cowan, Douglas E. *Sacred Space: The Quest for Transcendence in Science Fiction Film and Television*. Waco, Tex.: Baylor University Press, 2010.

Crawford, Ian A. "Interstellar Travel: A Review." In Zuckerman and Hart, *Extraterrestrials*, 50–69.

Crowe, Michael J. *The Extraterrestrial Life Debate, 1750–1900: The Idea of a Plurality of Worlds from Kant to Lowell*. New York: Cambridge University Press, 1986.

Darling, David, and Dirk Schulze-Makuch. *Megacatastrophes! Nine Strange Ways the World Could End*. Oxford, U.K.: Oneworld, 2012.

Davidson, Keay. *Carl Sagan: A Life*. New York: Wiley, 1999.

Davies, Paul. *Are We Alone? Philosophical Implications of the Discovery of Extraterrestrial Life*. New York: Basic Books, 1995.

———. *The Eerie Silence: Renewing Our Search for Alien Intelligence*. New York: Houghton Mifflin Harcourt, 2010.

———. "Life, Mind, and Culture as Fundamental Properties of the Universe." In Dick and Lupisella, *Cosmos and Culture*, 383–97.

———. "Transformations in Spirituality and Religion." In Tough, *When SETI Succeeds*, 51–52.

Dean, Jodi. *Aliens in America: Conspiracy Cultures from Outerspace to Cyberspace*. Ithaca: Cornell University Press, 1998.

Del Rey, Lester. *The World of Science Fiction, 1926–1976*. New York: Garland, 1980.

Denning, Kathryn. "Unpacking the Great Transmission Debate." In Vakoch, *Communication with Extraterrestrial Intelligence*, 237–52.

DeVore, Edna. "Closing In on E.T.'s Home." *SETI Institute Explorer*, 2012, 18–20.

———. "Making Science Education Exciting." *SETI Institute Explorer* 5, no. 1 (2008): 15–16.

DeVore, Edna, Jill Tarter, Jane Fisher, Kathleen O'Sullivan, Yvonne Pendleton, Sam Taylor, and Margaret Burke. "Educating the Next Generation of SETI Scientists: Voyages through Time." *Acta Astronautica* 53, no. 4–10 (August–November 2003): 841–46.

Dick, Steven J. "Bringing Culture to Cosmos: The Postbiological Universe." In Dick and Lupisella, *Cosmos and Culture*, 463–87.

———. "Cosmic Evolution." In Dick and Lupisella, *Cosmos and Culture*, 25–59.

———. *Life on Other Worlds: The 20th-Century Extraterrestrial Life Debate*. New York: Cambridge University Press, 1998.

———. *Plurality of Worlds: The Origins of the Extraterrestrial Life Debate from Democritus to Kant*. New York: Cambridge University Press, 1982.

———. "The Search for Extraterrestrial Intelligence and the NASA High Resolution Microwave Survey (HRMS): Historical Perspectives." *Space Science Reviews* 64 (1993): 93–139.

Dick, Steven J., and Mark J. Lupisella, eds. *Cosmos and Culture: Cultural Evolution in a Cosmic Context*. Washington, D.C.: NASA, 2010.

Dick, Steven J., and James E. Strick. *The Living Universe: NASA and the Development of Astrobiology*. New Brunswick, N.J.: Rutgers University Press, 2005.

Dickson, Paul. *Sputnik: Shock of the Century*. New York: Walker, 2007.

Disch, Thomas. *The Dreams Our Stuff Is Made Of: How Science Fiction Conquered the World*. New York: The Free Press, 1998.

Drake, Frank. "Intelligent Life in Other Parts of the Universe." In *The Earth in Space*, edited by Hugh Odishaw, 308–16. New York: Basic Books, 1967.

———. *Intelligent Life in Space*. New York: Macmillan, 1962.

———. "Methods of Communication: Message Content, Search Strategy, Interstellar Travel." In Ponnamperuma and Cameron, *Interstellar Communication*, 118–39.

———. "On Hands and Knees in Search of Elysium." *Technology Review* 78 (June 1976): 22–29.

Drake, Frank, and Dava Sobel. *Is Anyone Out There? The Scientific Search for Extraterrestrial Intelligence*. New York: Delacorte Press, 1992.

Dreier, Casey. "Breaking Sisyphus' Curse." *Planetary Report* 34, no. 2 (June Solstice 2014): 22–23.

Dyson, Freeman J. "Search for Artificial Stellar Sources of Infrared Radiation." In Cameron, *Interstellar Communication*, 111–14.

Easterbrook, Gregg. "Are We Alone?" *Atlantic Monthly*, August 1988, 25–38.

———. "We're All Gonna Die." *Wired*, July 2003, 151–57.

Eberhart, Jonathan. "Voyager's Message: Record Reviews." *Science News*, October 1, 1977, 211, 221.

———. "The World on a Record." *Science News*, August 20, 1977, 124–25.

Efron, Noah J. "Myth 9: That Christianity Gave Birth to Modern Science." In Numbers, *Galileo Goes to Jail*, 79–89.

Ehrman, Jerry. "'WOW!'—A Tantalizing Candidate." In Shuch, *Searching for Extraterrestrial Intelligence*, 47–63.

Eiseley, Loren. *Darwin's Century: Evolution and the Men Who Discovered It*. Garden City, N.Y.: Doubleday, 1958.

Eisenberg, Anne. "Philip Morrison—A Profile." *Physics Today*, August 1982, 36–41

Ekers, Ronald D., D. Kent Cullers, John Billingham, and Louis K. Scheffer, eds. *SETI 2020: A Roadmap for the Search for Extraterrestrial Intelligence*. Mountain View, Calif.: SETI Press, 2002.

Extraterrestrial Intelligence Research. Hearings before the Committee on Science and Technology, Subcommittee on Space Science and Applications, U.S. House of Representatives, September 19, 20, 1978. (Reprint from the collection of the University of Michigan Library)

Farmelo, Graham, ed. *It Must Be Beautiful: Great Equations of Modern Science*. London: Granta Books, 2002.

Finney, Ben, and Jerry Bentley. "A Tale of Two Analogues: Learning at a Distance from the Ancient Greeks and Maya and the Problem of Deciphering Extraterrestrial Radio Transmissions." *Acta Astronautica* 42, nos. 10–12 (May–June 1998): 691–96.

Finney, B., L. Finney, and V. Lytkin. "Tsiolkovsky and Extraterrestrial Intelligence." *Acta Astronautica* 46, nos. 10–12 (2000): 745–49.

"First Soviet-American Conference on Communication with Extraterrestrial Intelligence." *Spaceflight* 14, no. 1 (1972): 18–19.

Fletcher, James C. "The Space Program—A Social Enigma." In Frye, *Impact of Space Exploration,* 41–47.

Frosch, Robert A. Introduction. In Billingham, *Life in the Universe,* xiii–xv.

Frye, William E., ed. *Impact of Space Exploration on Society.* AAS Science and Technology Series, vol. 8. Washington, D.C.: American Astronautical Society, 1966.

Garber, Stephen J. "Searching for Good Science: The Cancellation of NASA's SETI Program." *Journal of the British Interplanetary Society* 52 (1999): 3–12.

Gardner, Martin. *Fads and Fallacies in the Name of Science.* 2nd ed. New York: Dover Publications, 1957.

Geirland, John. "The Quiet Zone." *Wired,* February 2004, 106–8, 140–42.

Genta, Giancarlo. *Lonely Minds in the Universe.* New York: Copernicus Books, 2007.

Geraci, Dan. "Annual Report to Our Members." *Planetary Report* 33, no. 2 (June 2013): 22–23.

Gilbert, James. *Redeeming Culture: American Religion in an Age of Science.* Chicago: University of Chicago Press, 1997.

Goldsmith, Donald, ed. *The Quest for Extraterrestrial Life: A Book of Readings.* Mill Valley, Calif.: University Science Books, 1980.

Graham, Loren R., ed. *Science and the Soviet Social Order.* Cambridge, Mass.: Harvard University Press, 1990.

Gray, Robert H. *The Elusive Wow: Searching for Extraterrestrial Intelligence.* Chicago: Palmer Square Press, 2012.

———. "The Fermi Paradox Is Neither Fermi's Nor a Paradox." *Astrobiology* 15, no. 3 (March 2015): 195–99.

Greenstein, Jesse, ed. *Astronomy and Astrophysics for the 1970's. Vol. 1: Report of the Astronomy Survey Committee.* Washington, D.C.: National Academy of Sciences, 1972.

Grinspoon, David. *Lonely Planets: The Natural Philosophy of Alien Life.* New York: HarperCollins, 2003.

———. "SETI and the Science Wars." *Astronomy* 28, no. 5 (May 2000): 52–57.

Guthke, Karl S. *The Last Frontier: Imagining Other Worlds, from the Copernican Revolution to Modern Science Fiction.* Translated by Helen Atkins. Ithaca: Cornell University Press, 1990.

Hagemeister, Michael. "Russian Cosmism in the 1920s and Today." In Rosenthal, *The Occult in Russian and Soviet Culture,* 185–202.

Hague, Angela. "Before Abduction: The Contactee Narrative of the 1950s." *Journal of Popular Culture* 44, no. 3 (June 2011): 439–54.

Hanrahan, James Stephen, ed. *The Search for Extraterrestrial Life. Advances in the Astronautical Sciences, vol. 22. Proceedings of the Twelfth Annual Meeting of the American Astronautical Society, May 23–25, 1966.* Sun Valley, Calif.: Scholarly Publications, 1967.

Harrison, Albert A. *After Contact: The Human Response to Extraterrestrial Life*. New York: Plenum Press, 1997.

———. "After Contact—Then What?" In Shuch, *Searching for Extraterrestrial Intelligence*, 497–514.

———. "The Science and Politics of SETI: How to Succeed in an Era of Make-Believe History and Pseudoscience." In Vakoch and Harrison, *Civilizations beyond Earth*, 141–55.

Hart, Michael. "An Explanation for the Absence of Extraterrestrials on Earth." *Quarterly Journal of the Royal Astronomical Society* 16 (1975): 118–35

Herzing, Denise L. "SETI Meets a Social Intelligence: Dolphins as a Model for Real-time Interaction and Communication with a Sentient Species." *Acta Astronautica* 67, nos. 11–12 (December 2010): 1451–54.

Hicks, R. D. *Stoic and Epicurean*. New York: Russell and Russell, 1962.

Hingley, Ronald. *Joseph Stalin: Man and Legend*. New York: McGraw-Hill, 1974.

Holquist, Michael. "The Philosophical Bases of Soviet Space Exploration." *Key Reporter* 51, no. 2 (Winter 1985–86): 2–4.

Horvath, Joan C. "Mixed Signals for ARECIBO." *Ad Astra*, Spring 2008, 47–49.

Hoyle, Fred. "Can We Learn from Other Planets?" *Saturday Review*, November 7, 1964, 63–67.

Hoyt, William Graves. *Lowell and Mars*. Tucson: University of Arizona Press, 1976.

Kaplan, S. A., ed. *Extraterrestrial Civilizations: Problems of Interstellar Communication*. Translated from the Russian. Jerusalem: Israel Program for Scientific Translations, 1971.

Kardashev, Nikolai. "Transmission of Information by Extraterrestrial Civilizations." In Tovmasyan, *Extraterrestrial Civilizations*, 19–29.

Kaufman, Marc. *First Contact*. New York: Simon and Schuster, 2011.

Kerr, Richard A. "NASA's Search for ETs Hits a Snag on Earth." *Science*, July 20, 1990, 249–50.

———. "SETI Faces Uncertainty on Earth and in the Stars." *Science*, October 2, 1992, 27.

Kilgore, De Witt Douglas. *Astrofuturism: Science, Race, and Visions of Utopia in Space*. Philadelphia: University of Pennsylvania Press, 2003.

Koerner, David, and Simon LeVay. *Here Be Dragons: The Scientific Quest for Extraterrestrial Life*. New York: Oxford University Press, 2000.

Korpela, Eric. "Distributed Processing of SETI Data." In Shuch, *Searching for Extraterrestrial Intelligence*, 183–99.

Krieger, F. J. "Space Programs of the Soviet Union." In Frye, *Impact of Space Exploration*, 211–19.

Larson, Edward J. *Evolution: The Remarkable History of a Scientific Theory*. New York: Modern Library, 2004.

Lasser, David. *The Conquest of Space*. 1931. Reprint, Burlington, Ont., Canada: Apogee Books, 2002.

Leigh, Darren, and Paul Horowitz. "Millions and Billions of Channels: A History of the Harvard SETI Groups's Searches." In Shuch, *Searching for Extraterrestrial Intelligence*, 107–27.

Leslie, Desmond, and George Adamski. *Flying Saucers Have Landed.* New York: The British Book Center, 1953.

Lieberman, Robbie. *The Strangest Dream: Communism, Anticommunism, and the U.S. Peace Movement, 1945–1963.* Syracuse, N.Y.: Syracuse University Press, 2000.

Lilly, John C., and Francis Jeffrey. *John Lilly, So Far.* Los Angeles: J. B. Tarcher, 1990.

Lindberg, David C. "Myth 1: That the Rise of Christianity Was Responsible for the Demise of Ancient Science." In Numbers, *Galileo Goes to Jail,* 8–18.

Lowell, Percival. *Mars as the Abode of Life.* New York: Macmillan, 1908.

Lupisella, "Cosmocultural Evolution." In Dick and Lupisella, *Cosmos and Culture,* 321–59.

Lytkin, Vladimir, Ben Finney, and Liudmila Alepko. "Tsiolkovsky, Russian Cosmism and Extraterrestrial Intelligence." *Quarterly Journal of the Royal Astronomical Society* 36 (December 1995): 369–76.

Masters, Dexter, and Katherine Way, eds. *One World or None.* 1946. Freeport, N.Y.: Books for Libraries Press, 1972.

McCurdy, Howard E. *Space and the American Imagination.* Washington, D.C.: Smithsonian Institution Press, 1997.

McDonough, Thomas R. *The Search for Extraterrestrial Intelligence: Listening for Life in the Cosmos.* New York: Wiley, 1987.

Michaud, Michael. *Contact with Alien Civilizations.* New York: Copernicus Books, 2007.

———. "A Unique Moment in Human History." In Bova and Preiss, *Are We Alone in the Cosmos,* 265–84.

Moffitt, John F. *Picturing Extraterrestrials: Alien Images in Modern Mass Culture.* Amherst, N.Y.: Prometheus Books, 2003.

Morrison, Philip. "If the Bomb Gets Out of Hand." In Masters and Way, *One World or None,* 1–6.

———. "Interstellar Communication." In Goldsmith, *The Quest for Extraterrestrial Life,* 122–31.

Morrison, Philip, John Billingham, and John Wolfe. "The Search for Extraterrestrial Intelligence." In Billingham and Pesek, *Communication with Extraterrestrial Intelligence,* 11–31.

———. *The Search for Extraterrestrial Intelligence: SETI.* Washington, D.C.: NASA, Scientific and Technical Information Office, 1977.

Morrison, Philip, and Kosta Tsipis. *Reason Enough to Hope: America and the World of the 21st Century.* Cambridge, Mass.: MIT Press, 1998.

Morton, Oliver. "A Mirror in the Sky: The Drake Equation." In Farmelo, *It Must Be Beautiful,* 171–92.

Mumford, Lewis. *The Story of Utopias.* New York: Boni and Liveright, 1922.

Murray, Brian. *H. G. Wells.* New York: Continuum, 1990.

National Research Council, Astronomy Survey Committee. *Astronomy and Astrophysics for the 1980's. Vol. 1: Report of the Astronomy Survey Committee.* Edited by George G. Field. Washington, D.C.: National Academy Press, 1982.

Numbers, Ronald L., ed. *Galileo Goes to Jail and Other Myths about Science and Religion*. Cambridge, Mass.: Harvard University Press, 2009.

Oliver, Bernard M. "NASA's Role in Man's Future." In Oliver, *Selected Papers*, 5–11.

———. "Reflections: The Microwave Observing Project Is Prologue." In Oliver, *Selected Papers*, 70–71.

———. *The Selected Papers of Bernard M. Oliver*. Palo Alto, Calif.: Hewlett-Packard, 1997.

———. "Technical Considerations in Interstellar Communication." In Ponnamperuma and Cameron, *Interstellar Communication*, 140–67.

Oliver, Bernard, and John Billingham. *Project Cyclops: A Design Study of a System for Detecting Extraterrestrial Life*. Moffett Field, Calif.: NASA/Ames Research Center, 1971.

O'Meara, Thomas F. *Vast Universe: Extraterrestrials and Christian Revelation*. Collegeville, Minn.: Liturgical Press, 2012.

Osiatynski, Wiktor, ed. *Contrasts: Soviet and American Thinkers Discuss the Future*. Translated by Ewa Woydyllo. New York: Macmillan, 1984.

Papagiannis, Michael. "The Fermi Paradox and Alternative Search Strategies: Introduction." In Papagiannis, *The Search for Extraterrestrial Life*, 437–42.

———. "A Historical Introduction to the Search for Extraterrestrial Life." In Papagiannis, *The Search for Extraterrestrial Life*, 5–11.

———. "The Hunt Is On." In Bova and Preiss, *Are We Alone in the Cosmos?* 205–21.

———, ed. *The Search for Extraterrestrial Life: Recent Developments: Proceedings of the 112th Symposium of the International Astronomical Union Held at Boston University, June 18–21, 1984*. Dordrecht, Holland: D. Reidel, 1985.

———, ed. *Strategies for the Search for Life in the Universe: A Joint Session of Commissions 16, 40, and 44, Held in Montreal, Canada, during the IAU General Assembly, 15 and 16 August, 1979*. Dordrecht, Holland: D. Reidel, 1980.

Pearson, J. P. T. "Extraterrestrial Intelligent Life and Interstellar Communication: An Informal Discussion." In Cameron, *Interstellar Communication*, 287–93.

Peebles, Curtis. *Watch the Skies: A Chronicle of the Flying Saucer Myth*. Washington, D.C.: Smithsonian Institution Press, 1994.

Pettinico, George. "American Attitudes about Life beyond Earth: Beliefs, Concerns, and the Role of Education and Religion in Shaping Public Perceptions." In Vakoch and Harrison *Civilizations beyond Earth*, 102–17.

Pierson, Tom. "Birth of the SETI Institute." *SETI Institute Explorer*, 2010, 4–6.

Pitta, Julie. "The Forbes Four Hundred and the Little Green Men." *Forbes*, October 17, 1994, 49.

Plank, Robert. *The Emotional Significance of Imaginary Beings*. Springfield, Ill.: Charles C. Thomas, 1968.

Ponnamperuma, Cyril, and A. G. W. Cameron, eds. *Interstellar Communication: Scientific Perspectives*. New York: Houghton Mifflin, 1974.

Poundstone, William. *Carl Sagan: A Life in the Cosmos*. New York: Henry Holt, 1999.

Proposed Studies on the Implications of Peaceful Space Activities for Human Affairs.

Report prepared by the Committee on Science and Astronautics, U.S. House of Representatives, April 18, 1961.

Purcell, Edward. "Radioastronomy and Communication through Space." In Cameron, *Interstellar Communication*, 121–43.

Raymo, Chet. *Skeptics and True Believers: The Exhilarating Connection between Science and Religion.* New York: Walker and Company, 1998.

Rees, Martin. *Our Final Hour: A Scientist's Warning: How Terror, Error, and Environmental Disaster Threaten Humankind's Future in This Century—on Earth and Beyond.* New York: Basic Books, 2003.

Regis, Edward, ed. *Extraterrestrials: Science and Alien Intelligence.* New York: Cambridge University Press, 1985.

Rheingold, Howard. "You Got the Power." *Wired,* August 2000, 176–84.

Roland, Paul. *Investigating the Unexplained.* New York: Berkley Books, 2001.

Rood, Robert T., and James S. Trefil. *Are We Alone? The Possibility of Extraterrestrial Civilizations.* New York: Scribner, 1981.

Rosenthal, Bernice Glatzer, ed. *The Occult in Russian and Soviet Culture.* Ithaca: Cornell University Press, 1997.

Ross, Monte, and Stuart Kingsley. "Optical SETI: Moving toward the Light." In Shuch, *Searching for Extraterrestrial Intelligence*, 147–82.

Ruse, Michael. "Is Rape Wrong on Andromeda? An Introduction to Extraterrestrial Evolution, Science, and Morality." In Regis, *Extraterrestrials*, 43–78.

Russell, Dale A. "Speculations on the Evolution of Intelligence in Multicellular Organisms." In Billingham, *Life in the Universe*, 259–75.

Sagan, Carl. *Billions and Billions: Thoughts on Life and Death at the Brink of the Millennium.* New York: Random House, 1997.

———, ed. *Communication with Extraterrestrial Intelligence (CETI).* Cambridge, Mass.: MIT Press, 1973.

———. *Contact.* New York: Pocket Books, 1986.

———. *The Cosmic Connection: An Extraterrestrial Perspective.* Garden City, N.Y.: Anchor Books, 1973.

———. *Cosmos.* New York: Random House, 1980.

———. "Direct Contact among Galactic Civilizations by Relativistic Interstellar Spaceflight." In Goldsmith, *The Quest for Extraterrestrial Life*, 205–13.

———. *The Dragons of Eden.* New York: Ballantine Books, 1977.

Sagan, Carl, and Frank Drake. "Extraterrestrial Intelligence: An International Petition." *Science,* October 29, 1982, 218.

———. "The Search for Extraterrestrial Intelligence." *Scientific American,* May 1975, 80–89.

Sagan, Carl, Frank Drake, Ann Druyan, Timothy Ferris, Jon Lomberg, and Linda Salzman Sagan. *Murmurs of Earth: The Voyager Interstellar Record.* New York: Random House, 1978.

Sagan, Carl, Linda Salzman Sagan, and Frank Drake. "A Message from Earth." In Goldsmith, *The Quest for Extraterrestrial Life*, 274–77.

Schewe, Phillip F. *Maverick Genius: The Pioneering Odyssey of Freeman Dyson*. New York: Thomas Dunne Books, 2013.

Shank, Michael H. "Myth 2: That the Medieval Christian Church Suppressed the Growth of Science." In Numbers, *Galileo Goes to Jail*, 19–27.

Shklovskii, I. S. *Five Billion Vodka Bottles to the Moon: Tales of a Soviet Scientist*. Translated and adapted by Mary Fleming Zirin and Harold Zirin. New York: Norton, 1991.

———. "Is Communication Possible with Intelligent Beings on Other Planets?" In Cameron, *Interstellar Communication*, 5–16.

Shklovskii, I. S., and Carl Sagan. *Intelligent Life in the Universe*. New York: Dell, 1966.

Short, Robert. *The Gospel from Outer Space*. New York: Harper and Row, 1983.

Shostak, Seth. "Astrobiology Growing Pains." *SETI Institute Explorer*, 2006, 14–15.

———. *Confessions of an Alien Hunter: A Scientist's Search for Extraterrestrial Intelligence*. Washington, D.C.: National Geographic Society, 2009.

———. *Sharing the Universe: Perspectives on Extraterrestrial Life*. Berkeley, Calif.: Berkeley Hills Books, 1998.

———. "Tuning In to Science." *SETI Institute Explorer*, 2012, 15–16.

Shuch, H. Paul. "A Half-Century of SETI Science." In Shuch, *Searching for Extraterrestrial Intelligence*, 5–11.

———. "Project Argus: Pursuing Amateur All-Sky SETI." In Shuch, *Searching for Extraterrestrial Intelligence*, 201–25.

———, ed. *Searching for Extraterrestrial Intelligence: SETI Past, Present, and Future*. Berlin: Springer, 2011.

Sieveking, Paul. "Forteana." *New Statesman and Society*, October 11, 1991, 31.

Simpson, George Gaylord. "The Nonprevalence of Humanoids." In Goldsmith, *The Quest for Extraterrestrial Life*, 214–21.

Smith, Marcia S. *Possibility of Intelligent Life Elsewhere in the Universe: Report Prepared for the Committee on Science and Technology, U.S. House of Representatives, Ninety-fifth Congress*. Washington, D.C.: U.S. Government Printing Office, 1977.

Spangenburg, Ray, and Kit Moser. *Carl Sagan: A Biography*. Amherst, N.Y.: Prometheus Books, 2009.

Squeri, Lawrence. "When ET Calls: SETI Is Ready." *Journal of Popular Culture* 37, no. 3 (February 2004): 478–96.

Strauss, David. *Percival Lowell: The Culture and Science of a Boston Brahmin*. Cambridge, Mass.: Harvard University Press, 2001.

Struve, Otto. "Astronomers in Turmoil." *Physics Today*, September 1960, 18–23.

Sullivan, Walter. *We Are Not Alone: The Search for Intelligent Life on Other Worlds*. Rev. ed. New York: McGraw-Hill, 1966.

Sullivan, Woodruff T., III. "SETI Conference at Tallinn." *Sky and Telescope* 63 (April 1982): 350–53.

Swift, David W. *SETI Pioneers: Scientists Talk about Their Search for Extraterrestrial Intelligence*. Tucson: University of Arizona Press, 1990.

Tarter, Donald E. "Can SETI Fulfill the Value Agenda of Cultural Anthropology?" In Vakoch and Harrison, *Civilizations beyond Earth*, 87–101.

Tarter, Jill. "Exoplanets, Extremophiles, and the Search for Extraterrestrial Intelligence." In Vakoch, *Communication with Extraterrestrial Intelligence*, 3–18.

———. "SETI and the Religions of Extraterrestrials." *Free Inquiry* 20, no. 3 (Summer 2000): 34–35.

Terkel, Studs. *The Good War: An Oral History of World War Two*. New York: Pantheon, 1984.

Tipler, Frank J. "Additional Remarks on Extraterrestrial Intelligence." *Quarterly Journal of the Royal Astronomical Society* 22 (1981): 279–92.

———. "A Brief History of the Extraterrestrial Intelligence Concept." *Quarterly Journal of the Royal Astronomical Society* 22 (1981): 133–45.

———. "Extraterrestrial Intelligent Beings Do Not Exist." *Quarterly Journal of the Royal Astronomical Society* 21 (1980): 267–81.

Tosh, John, ed. *Historians on History*. Harlow, U.K.: Pearson, 2000.

Tough, Allen. "An Extraordinary Event." In Tough, *When SETI Succeeds*, 1–6.

———, ed. *When SETI Succeeds: The Impact of High-Information Contact*. Bellevue, Wash.: Foundation for the Future, 2000.

Tovmasyan, G. M., ed. *Extraterrestrial Civilizations: Proceedings of the First All-Union Conference on Extraterrestrial Civilizations and Interstellar Communication, Byurakan, 20–23 May 1964*. Jerusalem: Israel Program for Scientific Translations, 1967.

Townes, C. H., and R. N. Schwartz. "Interstellar and Interplanetary Communication by Optical Masers." *Nature* 190, no. 4772 (April 15, 1961): 205–8.

U.S. Congress, Senate. *Congressional Record*. 97th Cong., 1st sess., July 30, 1981: Senator Proxmire on SETI. 18633–35.

———. 103rd Cong., 1st sess., September 22–23, 1993. Debate on The Search for Extraterrestrial Intelligence Program. 21713–14, 21964–68.

Vakoch, Douglas. "Asymmetry in Active SETI: A Case for Transmissions from Earth." *Acta Astronautica* 68, nos. 3–4 (February–March 2011): 476–88.

———, ed. *Communication with Extraterrestrial Intelligence*. Albany: SUNY Press, 2011.

———. "Integrating Active and Passive SETI Programs: Prerequisites for Multigenerational Research." In Vacoch, *Communication with Extraterrestrial Intelligence*, 253–78.

———. "Reactions to Receipt of a Message from Extraterrestrial Intelligence: A Cross-Cultural Empirical Study." *Acta Astronautica* 46, nos. 10–12 (2000): 737–44.

———. "What's Past Is Prologue: Future Messages of Cosmic Evolution." In Shuch, *Searching for Extraterrestrial Intelligence*, 373–98.

Vakoch, Douglas A., and Albert A. Harrison, eds. *Civilizations beyond Earth: Extraterrestrial Life and Society*. New York: Berghahn Books, 2011.

Verschuur, Gerrit L. "If We Are Alone, What on Earth Are We Doing?" *Sky and Telescope* 78 (November 1989): 452.

———. *The Invisible Universe: The Story of Radio Astronomy*. 2nd ed. New York: Springer, 2010.

———. *Is Anyone Out There? Personal Adventures in Search for Extraterrestrial Intelligence*. Charleston, N.C.: CreateSpace Independent Publishing, 2015.

———. "A Search for Narrow Band 21-cm Wavelength Signals from Ten Nearby Stars." *Icarus* 19 (1973): 329–40.

Viewing, David. "Directly Interacting Extraterrestrial Communities." *Journal of the British Interplanetary Society* 28 (1975): 735–44.

Von Hoerner, Sebastian. "The General Limits of Space Travel." In Goldsmith, *The Quest for Extraterrestrial Life*, 197–204.

Ward, Peter, and Donald Brownlee. *Rare Earth: Why Complex Life Is Uncommon in the Universe*. New York: Springer-Verlag, 2000.

Wattenberg, Ben J. *Fewer: How the New Demography of Depopulation Will Shape Our Future*. Chicago: Ivan Dee, 2004.

Webb, Stephen. *If the Universe Is Teeming with Aliens . . . Where Is Everybody?* New York: Copernicus Books, 2002.

———. "Pondering the Fermi Paradox." In Shuch, *Searching for Extraterrestrial Intelligence*, 305–21.

White, Frank. *The SETI Factor: How the Search for Extraterrestrial Intelligence Is Changing Our View of the Universe and Ourselves*. New York: Walker and Company, 1990.

Wilkinson, David. *Science, Religion, and the Search for Extraterrestrial Intelligence*. Oxford, U.K.: Oxford University Press, 2013.

Wolfe, John H. "On the Question of Interstellar Travel." In Papagiannis, *The Search for Extraterrestrial Life*, 449–54.

Wukelic, George, ed. *Handbook of Soviet Space-Science Research*. New York: Gordon and Breach, 1968.

Young, George M. *The Russian Cosmists: The Esoteric Futurism of Nikolai Fedorov and His Followers*. New York: Oxford University Press, 2012.

Zaitsev, Alexander. "METI: Messaging to Extraterrestrial Intelligence." In Shuch, *Searching for Extraterrestrial Intelligence*, 399–428.

Zuckerman, Ben, and Michael H. Hart, eds. *Extraterrestrials: Where Are They?* 2nd ed. Cambridge: Cambridge University Press, 1995.

INDEX

LAWRENCE SQUERI is professor emeritus of history at East Stroudsburg University of Pennsylvania.